In the Shadow of the Seawall

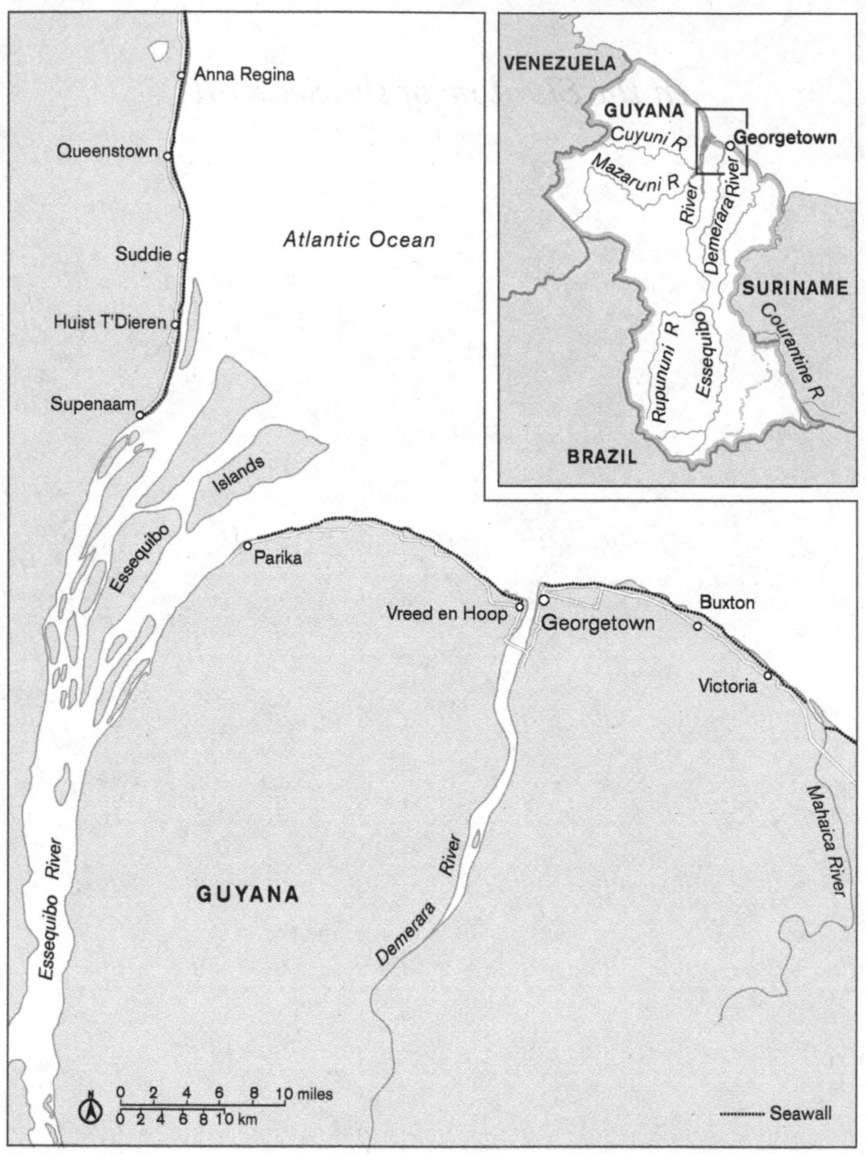
VENEZUELA
GUYANA
Cuyuni R
Georgetown
Mazaruni R
River
Demerara River
SURINAME
Courantine R
Rupununi R
Essequibo
BRAZIL
Anna Regina
Queenstown
Atlantic Ocean
Suddie
Huist T'Dieren
Supenaam
Essequibo Islands
Parika
Vreed en Hoop
Georgetown
Buxton
Victoria
Mahaica River
Demerara River
GUYANA
Essequibo River
0 2 4 6 8 10 miles
0 2 4 6 8 10 km
Seawall

Map 1. Guyana

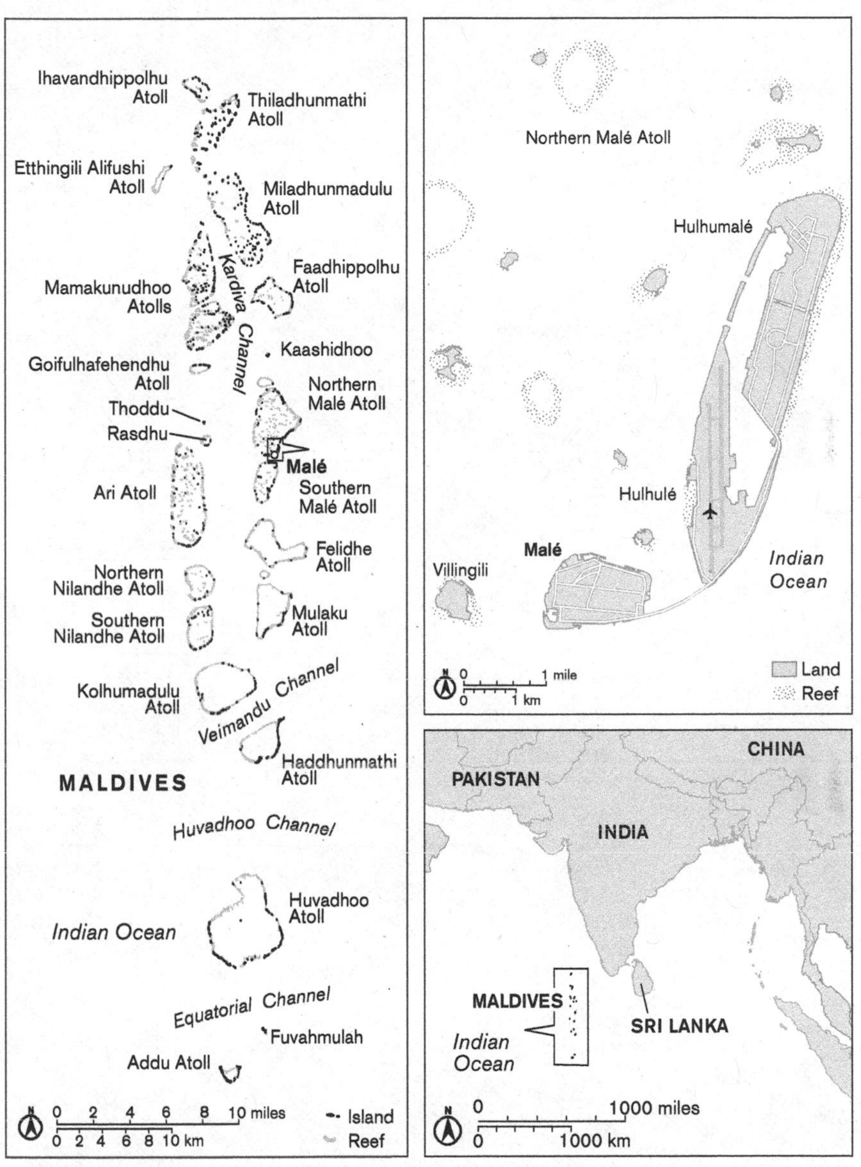

Map 2. The Maldives

In the Shadow of the Seawall

COASTAL INJUSTICE AND THE DILEMMA OF PLACEKEEPING

Summer Gray

UNIVERSITY OF CALIFORNIA PRESS

University of California Press
Oakland, California

Library of Congress Cataloging-in-Publication Data

Names: Gray, Summer, author.
Title: In the shadow of the seawall : coastal injustice and the dilemma of placekeeping / Summer Gray.
Description: Oakland, California : University of California Press, [2023] | Includes bibliographical references and index.
Identifiers: LCCN 2023018065 (print) | LCCN 2023018066 (ebook) | ISBN 9780520392731 (cloth) | ISBN 9780520392748 (paperback) | ISBN 9780520392755 (epub)
Subjects: LCSH: Sea-walls—Guyana. | Sea-walls—Maldives. | Coast changes—Guyana. | Coast changes—Maldives. | Coastal zone management—Guyana. | Coastal zone management—Maldives. | Climate justice—Guyana. | Climate justice—Maldives.
Classification: LCC TC335 .G73 2023 (print) | LCC TC335 (ebook) | DDC 333.91/7095495—dc23/eng/20230503
LC record available at https://lccn.loc.gov/2023018065
LC ebook record available at https://lccn.loc.gov/2023018066

Manufactured in the United States of America

32 31 30 29 28 27 26 25 24 23
10 9 8 7 6 5 4 3 2 1

To the frontline communities who stay and fight

Contents

Acknowledgments

Overlapping gestures of kindness, encouragement, and love accompanied me in this journey, and I am forever grateful to those who made it possible. First and foremost, I would like to thank the frontline residents of Guyana and the Maldives who shared their stories with me. Their voices permeate this book and remind me of how important it is to slow down, listen, and engage with complicated matters that defy traditional frameworks and understandings. I am also grateful to members of the Guyana Mangrove Restoration Project and the Maldivian Democratic Party who supported me during my fieldwork through their own transformative work, connecting me to the many different places and people that embody life in the shadow of the seawall.

This book started as a dissertation in the Department of Sociology at the University of California, Santa Barbara (UCSB). It would not have been possible without the support of mentors who trusted that a project on seawalls would have something meaningful to say about society. I would like to express my deepest appreciation to the members of my dissertation committee. I am especially grateful to Kum-Kum Bhavnani, who took me under her wing and kept me on course as I navigated uncharted waters. I also want to thank Richard Appelbaum, whose work on globalization expanded my thinking and encouraged me to consider multiple cases, and Janet Walker, who strengthened my desire to challenge dominant narratives. For opening doors that might have otherwise been closed, I am extremely grateful to John Foran, whose commitment to radical social change pushed me to embrace and amplify the revolutionary potential of the climate justice movement.

At the University of California, Santa Cruz, I was fortunate to receive mentorship from Anna L. Tsing. I would like to express my deepest gratitude to Anna for supporting my work and attuning me to the more-than-human world. For reading and commenting on early drafts of my chapters, I want to thank members of Anna's fall 2015 seminar "Planetary Transitions: Critical Landscape Ecologies of the Anthropocene." I would also like to thank the Anthropology Department staff for welcoming me and providing administrative support and an office with a beautiful view of the redwood trees that weave throughout the campus.

The fieldwork for this book was supported by several small grants, including an award from the James D. Kline Fund for International Studies, a Graduate Research Mentorship Fellowship, a mini-grant from the John Andrew W. Mellon Foundation, and travel funds from the University of California President's Postdoctoral Fellowship Program (PPFP). My fieldwork was further supported by those who shared their homes and workplaces with me. I want to thank Annette Arjoon-Martins and Dave Martins for their support of my research in Guyana and for connecting me with many key figures in the environmental community. I also want to thank the members of the Guyana Mangrove Restoration Project who invited me into their daily lives and accompanied me on many trips up and down the coast, including Kene Moseley, Ranata Robinson, Luan Gooding, Tana Yussuff, and Winston Walcott. For supporting me with my research in the Maldives, I am grateful to Mohamed Aslam and members of the Land and Marine Environmental Resource Group and Riyan who helped arrange interviews and interisland travel during my visit, including Shifu Saeed and Aisha Abdulla. I would also like to acknowledge Jane Da Mosto, Masako Ishiguro, and Tony Sevold for their tremendous help in Venice, Japan, and the Netherlands.

Many conversations have taken place over the years that have helped build the ideas in this book. A special thanks to Orinn Pilkey, Bruce Caron, and Charles Lester for sharing my obsession with seawalls and creating a lively conversation to build on. For expanding the scope of environmental sociology and engaging with my work during the special session "Landscapes of Inequality" at the 2021 American Sociological Association Annual Meeting, I am grateful to Amanda McMillan Lequieu, Mimi Sheller, Hillary Angelo, and Rebecca Elliott. My ideas expanded to new dimensions thanks to Alenda Chang, who helped organize the "Future Tripping" symposium at UCSB. I am grateful to Barbara Endemaño Walker, Erin Khue Ninh, Verónica Castillo-Muñoz, and Karen Lunsford for commenting on early drafts of my work. I also want to thank the Headlands Center for the Arts and the Institute for Sustainability, Energy, and Environment at the University of Illinois, Urbana-Champaign, for inviting me to share early versions of this work.

The ideas in this book also grew from many opportunities to engage with interdisciplinary intellectual communities at the interface of climate justice and environmental justice. I had the pleasure of collaborating with an amazing team of people through the UC-CSU NXTerra Knowledge Action Network, including Sarah Jaquette Ray, Amanda Baugh, Beth Rose Middleton, Tori Derr, Nicole Seymour, and Daniel Fernandez. I gained firsthand knowledge of how people are fighting back from my collaborations with members of the Climate Justice Project and the Environmental Justice and Climate Justice Research Hub. I would like to thank ann-elise lewallen, Elvia Cruz-Garcia, and Emily Williams for their enthusiasm and support over the years. I would also like to thank Teresa Shewry and Peter Alagona for organizing the Mellon-Sawyer Seminar on Sea Change and inviting me to think through the possibilities of marine justice.

The final stages of this project benefited tremendously from the generous support of the University of California (UC). From 2015 to 2017, the PPFP made it possible for me to expand my fieldwork to Japan. During this time, I gained valuable insight from members of the PPFP community, many of whom commented on early presentations of my work during our writing retreats in Lake Arrowhead. I also want to express my gratitude to the entire staff at UC Press, including Stacy Eisenstark, Naomi Schneider, and Naja Pulliam Collins, for supporting me and providing a forum to share this project with the world. I am enormously grateful to the anonymous reviewers who improved my chapters by volunteering copious amounts of time and intellectual energy to comment on earlier drafts. The revision stage of this book was supported by a UC Regents' Junior Faculty Award that enabled me to take time away from teaching to concentrate on this project.

In the Environmental Studies Program at UCSB, I am fortunate to be surrounded by a remarkable group of humanists, social scientists, and natural scientists who share a deep concern over the future of our shared planet. I want to thank David Pellow for paving the way and guiding me through my early years as an assistant professor while also providing valuable feedback on my book proposal. I am grateful to the Environmental Studies Program staff, who helped me turn the institutional wheels to process travel reimbursements, student hires, and other purchases that allowed me to make the finishing touches on this project. I am also grateful to my imaginative student-scholars, including Jamie Chen, Luna Herschenfeld-Catalan, Inês Laborinho Schwartz, Sophia Lin, Claire Muñoz, Taylor Roe, and Ian Silberstein, for helping me transcribe interviews and gather literature for this project.

This project was also enriched by the support of friends and colleagues. Noah Zweig ensured that I always had someone to talk to and rely on during the early stages of the research. Corrie Grosse was a constant source of inspiration and wisdom. Elana Resnick, Rebecca Powers, and Heather Steffen lifted me up, reminded me to take breaks, and held my hand digitally throughout the pandemic. Zakiya Luna, and Dolores Inés Casillas helped me more than they probably realize by fostering an empowering writing community at UCSB. I also owe much of my own personal resilience to the transformative work and influence of Abigail Reyes and Aryeh Shell.

It is difficult to imagine myself on this path without my family's love and support. I am deeply indebted to my sister, Jessica, my mom, Kathy, my dad, George, my second mom, ZoeAnn, my brother-in-law, Jaime, and all of my aunts and uncles and nephews. Over the past ten years, my family cheered me on, drove countless hours to visit me, and made sure that I was cared for. Throughout this entire journey, Jason supported me wholeheartedly and served as my intellectual and emotional compass. My late grandparents, Ruth and Ernest Perea, gave me strength and conviction. Their memory and capacity to overcome adversity continues to guide me.

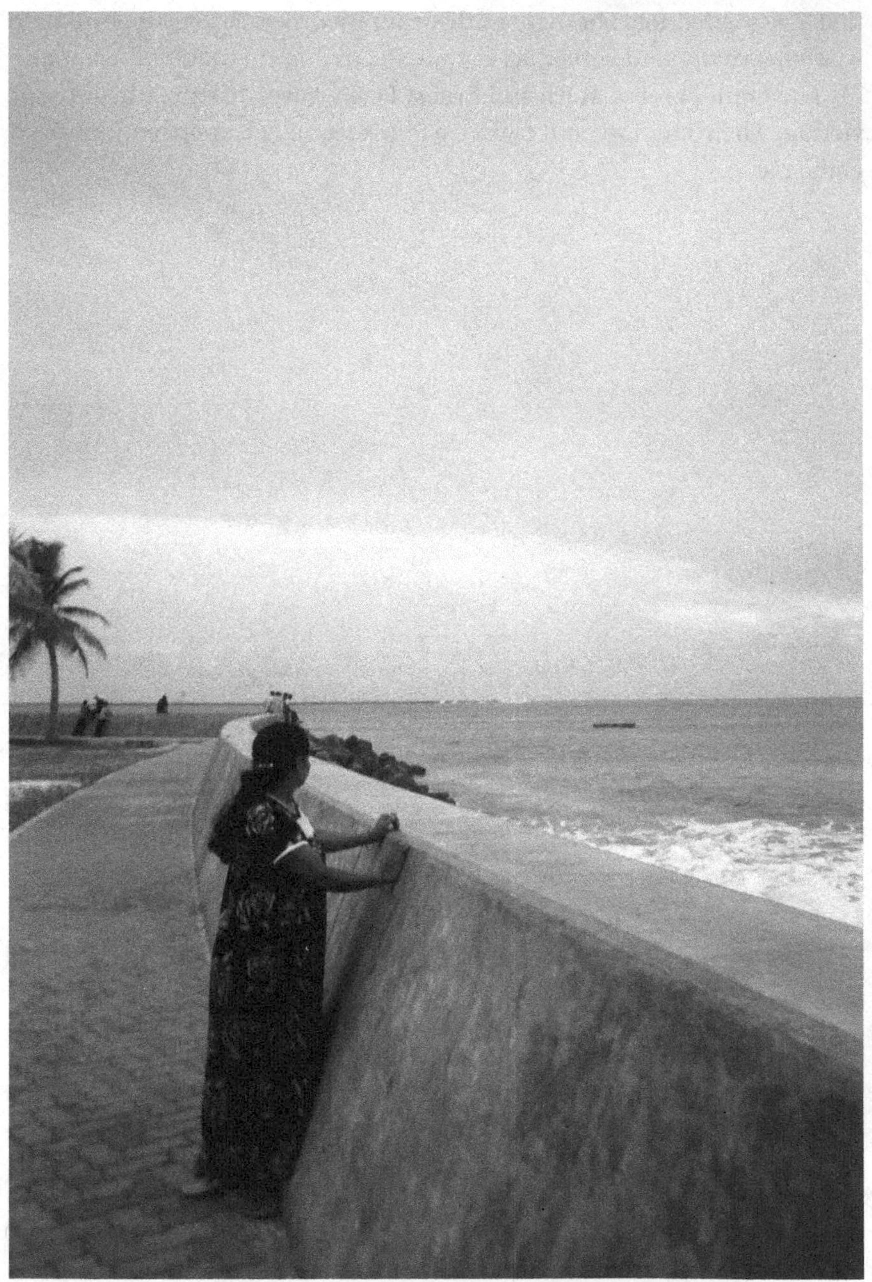

Figure 1. A woman pauses to reflect at a seawall in the Maldives, 2008.
Photo: Hani Amir

Introduction

SEAWALL ENTANGLEMENTS

I first realized that seawalls represent something complicated about modern life over a decade ago when I was observing community-based conservation efforts in Trinidad and Tobago. I was drawn to the plight of the leatherback sea turtle, a species endearingly known as Earth's last dinosaur. As one of the most ancient and migratory oceanic species, endangered leatherbacks have come to symbolize the ecological health of the planet.[1] In 2009, I traveled to the equatorial front lines of this struggle, where the circle of leatherback life is most frequently broken. On a moonlit beach in Matura, I monitored female turtles as they wriggled out from the Atlantic Ocean to perform their slow-motion birth rituals. Walking along the wave-battered shore at midnight, I trained my eyes to see in the dark while steering clear of the creatures lurking in the marshy puddles beyond the trees, silently repeating the mantra, "Never turn your back to the sea."

I would soon discover that a battle was encroaching. Erosion had prompted the construction of a seawall up the coast, the foundations of which were already protruding from the sandy shore like a baby's first teeth. Those in my team who understood coastal dynamics were quick to inform me that this nascent seawall was a bad omen. These structures were known to destabilize coastal ecosystems and set in motion a process of perpetual coastal squeeze, leading to a decline in beach sand and further disruptions for endangered sea turtles. The image of Sisyphus and his boulder stuck in a never-ending uphill battle came to my mind. This wall, at the interface of land and sea, humans and more-than-humans, was a sign of things to come, a monument of clashing desires for preservation in times of fierce and unprecedented change.

This initial observation led me to follow seawalls for several years. In writing this book, I floated between distant shores to places where complicated relationships of perseverance emerge in the shadows of coastal disruption and efforts to resist displacement. I placed a metaphorical finger against the edge of the sea to trace the contours of its altered state. I felt the scattered effects of global forces reshaping the shore with ever-increasing exploitation, sky-bound buildings, and shortsighted development projects. I also felt the resistance formed by humans who for different reasons have fought with concrete and stone to keep the sea at bay. I noticed that seawalls are always situated in social worlds, between groups of people, ecosystems, and human institutions where sentiments, memories, and livelihoods are invested.

Two seawall entanglements separated by a vast ocean captured my attention and illuminated the difficult realities of staying in place amid contexts of ecological crisis and political oppression. One is located in coastal Guyana, an alluvial swampland of rivers and mangroves. The other is in the islands of the Maldives, a glistening tropical paradise of sand and coral. These are the protagonists of this story; together, they uncover the dilemma of seawalls across divergent contexts of coastal disruption. Their ecologies represent two of the most common struggles among frontline communities: the shoreline that still bears the marks of colonization and plight of the low-lying island nation. These two contrasting images of coastal disruption complement one another and illuminate the tensions and contradictions of maintaining infrastructural assemblages across multiple and competing desires for permanence.

Together, Guyana and the Maldives underscore an uncomfortable reality—that staying in place on a warming planet is an inherently unequal struggle. Colonial dispossession, political oppression, land grabbing, and other forms of modern development have created an unfair burden of vulnerability. This continual and unrelenting burden consistently finds those who have contributed least to the problem and have fewer resources with which to cope with change, including capacities for protection. Over time, shorelines have increasingly become sites of erosion, toxic runoff, deadly weather, declining marine habitats, and battle zones of clashing public and private interests.

Nevertheless, the conviction to stay in place is a sentiment that echoes far and wide across the many front lines of climate change. I heard it loud and clear on the shores of Guyana, where women are planting green seawalls to maintain their livelihoods. I also heard it in the small island nation of the Maldives, where activists chant, "We have a right for home," and youth movements work to save the few remaining public beaches from washing away. In Guyana and the Maldives, and in many other nations at the front lines of climate change, staying in place is a dilemma—controversial, complex, and existential.

My term for this struggle is *placekeeping*, a process shaped by histories of oppression, attachments to place, and anticipation of loss. In its current usage among planners and artists, placekeeping implies the long-term maintenance of place and in particular the social and environmental qualities of place for the benefit of present and future generations.[2] According to the US Department of Arts and Culture, a grassroots action network, placekeeping "is not just preserving buildings but keeping the cultural memories associated with a locale alive, while supporting the ability of local people to maintain their way of life as they choose."[3] Processes of placekeeping become suddenly more meaningful in relation to the multiple crises of climate disruption.

I use placekeeping as an alternative to conventional notions of climate adaptation in order to cast critical light on how desires for permanence collide with oppression and experiences of loss. Placekeeping entails practices of maintenance that are embattled in local struggles to choose and define life-affirming pathways of existence in times of disruption. These pathways are being negotiated through complex entanglements of modern infrastructural systems and relationships of power that mirror long-standing struggles for democracy and sovereignty. In the context of climate change, and coastal disruption in particular, placekeeping is a dilemma that engages with shifting values of protection. These values are embedded in local place-based politics and global structures of inequality that give shape to the social, economic, and physical conditions of vulnerability. Access to protection is further shaped by practices of social and material dispossession at the interface of land and sea.

Placekeeping is about more than staying above water; it is a struggle to define what is worth saving. Conceptually, placekeeping situates the problem of climate adaptation within wider cultural and political contexts that create and sustain conditions of social and ecological precarity. As it stands, adaptation is a limited framework that operates on the assumption that shorelines are blank social canvases rather than landscapes shaped by conflicting desires and unequal relationships of power. In contrast, placekeeping turns our attention to historical and ongoing processes of vulnerability—including the making and remaking of place—while also accounting for the ways in which competing logics of adaptation enter into existing struggles to keep place and stay in place. For this reason, and others described throughout this book, I argue that adaptation needs to be reframed as a larger effort to maintain place, contextualized through wide-ranging struggles for justice across multiple scales and temporalities.

PLACING ADAPTATION

Despite the global nature of the present ecological crisis, the consequences continue to be experienced locally and unevenly, often in places where conflicting political and ecological meanings are most visible. For many frontline communities, including those in Guyana and the Maldives, climate mitigation will not come soon enough. Some form of adaptation will be necessary. This leads to a host of problems for which there are no easy solutions. When we zoom out to look at the problem of sea change through the widest lens possible, the solution can appear to be deceptively simple: move people and infrastructure landward to make room for the sea. Zoom in, though, and it becomes clear that every measure in response to sea change is embedded in existing values, economic limitations, and structural inequalities. At the root of the crisis is the problem of capitalism and land-use management, which breeds economic inequality, structural racism, unjust policies, and priorities that strain and challenge movements for justice.

As a field of study and practice, adaptation has been largely dedicated to understanding the anticipated losses associated with efforts to preserve human life and economic development. The focus has largely centered on the loss of place to *invisible* climatic forces and not to place-specific histories of dispossession, political oppression, and everyday efforts to survive. This narrow definition of loss leads to another loss—of attention to the many practices and efforts to stay in place that are in constant negotiation with ecological and political change. In places where adaptation is already under way, desires for permanence increasingly enter into longstanding confrontations with injustice. In the Global South, where choice is severely limited, surviving the future means negotiating with inequality and political oppression while also confronting the difficult realities of environments scarred by years of disruption.

Underlying the limited framework of climate adaptation are the conceptual drivers of resilience thinking. Adaptation discourse is channeled through a predominantly top-down, expert-driven concept of *resilience* that is framed to accommodate threats, not to prevent them.[4] In its most common usage, resilience refers to the capacity of systems to withstand severe conditions and absorb shocks and is guided by a widely abused metaphor of ecological equilibrium.[5] Some argue that the turn to resilience is a means of preserving modernist imaginaries and utopian visions of development and progress. As Danny MacKinnon and Kate Derickson observe, "Resilient spaces are precisely what capitalism needs—spaces that are periodically reinvented to meet the changing demands of capital accumulation in an increasingly globalized economy."[6] When used by planners, "resilience" reflects an idealism about social progress.[7] When applied to social systems, theories of resilience focus on sustaining what exists rather than tackling what exists as the problem itself.[8]

Conceptually, resilience is often framed at the scale of individual communities, failing to consider the larger systemic processes that have contributed to the conditions of vulnerability in the first place. Eija Meriläinen, who researches disaster governance, describes this problem as "scalar disconnect," where governments work to intervene on the scale of the city as a whole while shifting the responsibility to individual neighborhoods to manage their own risk.[9] This includes the relinquishment of state responsibilities to nongovernmental organizations (NGOs), a trend that often occurs after a major disaster through international disaster relief funding. Resilience planners often ask vulnerable populations to transform themselves in drastic ways that are not asked of "nonvulnerable" populations, burdening communities with sacrifices while shifting responsibility away from the state. For this reason, resilience is said to have a dark side.[10]

When framed within conventional resilience logics, adaptation is a process that tends to exacerbate inequality. Because conditions of vulnerability are often situated in a larger context of what Cedric Robinson identifies as *racial capitalism*, adaptation frameworks that wash over existing racial formations and spatial inequalities serve to heighten racial injustice. This was the case in post-Katrina New Orleans, where state-mandated resilience planning translated into increased police presence and evictions.[11] In relation to sea change, resilience planning has done more to secure the interests of wealthy white property owners.[12] Dean Hardy and colleagues at the front lines of sea change in Florida refer to this as a form of colorblind adaptation planning that ignores histories of racial injustice.[13] The main problem with resilience thinking is that it reduces unequal power relations to the singular problem of maintaining the status quo, which impedes efforts to enact social and environmental justice or solutions that involve progressive change.

The technical and policy-oriented focus of mainstream climate adaptation—including the economic, colonial, and patriarchal structures on which it relies—obscures the historical and social complexity of climate change, further reinforcing vulnerabilities to climate change. The professional field of resilience and adaptation planning is directly tied to funding institutions and agencies that are the progenitors of global inequality.[14] These are the same institutions that initiated the instruments of economic development and globalization, which include a range of assessment tools, private partnerships, and a marking network touted to create a "resilience dividend" in exchange for technical and managerial access to rebuild urban and developing areas.[15]

Driving this inattention to systemic relationships of injustice and vulnerability is a limited notion of temporality—a *future*-oriented focus on climate science and hazards that shifts temporality away from the past, absolving adaptation planners from historical responsibility. As a result, adaptation planning is fixated on innovative technology, green infrastructure, and market-driven solutions, losing sight of the causal connections and transformational changes necessary for eradicating the roots of climate vulnerability. Kasia Paprocki proposes *anticipatory ruination* as a conceptual tool for thinking about the ideological and material work of "future experts," who are both responding to and producing climate crises in developing nations.[16] Her insight, which takes inspiration from Ann Laura Stoler's *Imperial Debris*, suggests that anticipatory ruination in the context of climate change is a process driven mainly by experts and development agencies.

Part of the problem is that climate change is a process that is more often understood as a story of endings rather than one of beginnings. As such, it is widely assumed that climate change is a recent phenomenon with a clear trajectory. The Intergovernmental Panel on Climate Change, for instance, narrates climate change through a range of predicted future scenarios, querying the possible endpoints of global carbon emissions. The predicted outcomes, based on different configurations of social and political conditions, range from the benign to the catastrophic, the latter of which has become the more statistically likely scenario. However, climate disruption embodies a set of forces enacted over time, and the practice of locating its origins and the story of how humans came to disrupt the planet often bear more truth than the art of imagining its ends.

What conventional adaptation frameworks fail to consider is that there is value in understanding climate change as rooted in multiple beginnings. In fact, there are a number of scientifically plausible starting points for anthropogenic climate change, some dating as far back as the invention of fire and others pointing to the violence of colonization.[17] Dwelling on the origins of harm moves beyond what Jane Bennett refers to as "the project of blaming," allowing us to view disruption as a river of many failed dreams and wrongdoings.[18] From this perspective, vulnerability is not a static state or a singular condition; it entails a range of temporalities and practices that build momentum, collide, and overlap in waves that crash down on the coast. As Andreas Malm poignantly writes, "This is not a one-off event sometime in the future. It is utterly present and ongoing; it mixes with moments of old normality; it submerges some places and mercifully passes by others; it proceeds by creeping slowly and invading instantly. It has no clear end."[19]

This is not to suggest that all adaptation planning is oppressive or shortsighted. Some social and environmental activists are finding space to mobilize adaptation strategies that subvert entrenched power dynamics. The Movement Strategy Center, for example, embraces a form of resilience building they call "community-driven climate resilience." They define resilience differently from mainstream planners and academic researchers, emphasizing that building resilience "requires significant structural shifts to address the root causes of climate change, as well as comprehensive place-based innovations that increase social cohesion, localize food and energy systems, and advance democratic participation practices."[20] Resilience is perhaps best understood as "cultural work" that mirrors our collective grappling with climate change and a concept around which desires for stability and change are being negotiated.[21] Nonetheless, mainstream resilience discourses dominate the landscape of adaptation, suggesting a need for a more critical engagement with adaptation.

Without attention to *place* and historical practices of dispossession and land-use management, adaptation is nothing but an abstraction. Place can mean many things, but no matter how it is defined, it entails a coming together of people, ideas, values, physical objects, and ecologies that produce identities, livelihoods, politics, power dynamics, visions, vulnerabilities, and inequalities. While place is necessarily material, it is also imbued with value and meaning. Place grounds adaptation and resilience to existing struggles for social and environmental justice. *Placing* adaptation means venturing beyond conventional understandings of climate change vulnerability to consider the diverse relationships that exist across place and time, from colonial encounters to schemes for economic development. Not all places are created equal. There are important differences in how adaptation is conceptualized and operationalized in the Global North and in the Global South.[22]

Thomas Gieryn observes that recent trends in sociological theory, particularly globalization theory, have shifted place to the background in favor of detached notions of "space" and "flow."[23] Yet place remains central to sociological thought and is necessary for attending to some of the most pressing issues of our time, including intersecting issues of inequality and ecological disruption. This necessitates a shift back to what Gieryn terms "a place-sensitive sociology," where place is not confined to a separate area of study but is recognized as a fundamental epistemological thread of the sociological imagination. "Places are endlessly made," he writes, evoking Michel de Certeau's *The Practice of Everyday Life*, "not just when the powerful pursue their ambition through brick and mortar, not just when design professionals give form to function, but also when ordinary people extract from continuous and abstract space a bounded, identified, meaningful, named, and significant place."[24] The making of place is a never-ending process of social and material struggle.[25]

A place-sensitive approach to climate change expands our understanding of what is actually at risk of being lost. Rebecca Elliott, in her beautifully conceived article, "A Sociology of Loss," challenges conventional framings of adaptation.[26] In discussing the materiality of loss, she refers to a variety of losses, including the type in which "communities experience the disappearance of the land beneath their feet and, with it, the built and nonhuman environments that make social life possible and predictable[,] . . . disappearances wrought by shifting coastlines, denuded forests, storm-wrecked cities."[27] But she does not limit loss to its calculable, quantitative dimensions; she also draws attention to the qualitative, intangible qualities of loss, including loss by design and "how waterfronts, landmasses, cities, and villages become active sites of destructive transformation, with potentially unintended, contradictory, and unequal consequences."[28] This also includes attention to loss as abandonment, whose losses matter, what solidarities form in the process of loss, and "how place may be lost—or sacrificed—to protect or promote other cherished things and ideals."[29]

Placekeeping, as it is conceptualized in this book, provides a critical alternative to adaptation framings and bridges concerns posed by critical adaptation scholars and place-sensitive sociologists who understand the significance of place attachment and the anticipation of loss in contexts of climate disruption. It recognizes that projects that engage with the language of resilience and adaptation can hurt communities by ignoring historically entrenched injustices and ongoing place-based struggles.[30] Addressing climate change means addressing the roots of the harm, including processes of dispossession, abandonment, and injustice that underlie contemporary vulnerabilities.

Placekeeping builds on many efforts to define place through the empowerment of those who are its stewards, recognizing and embracing radical definitions of community resilience that challenge top-down politics. How adaptation practices come into contact with place-specific power dynamics is key to determining what effect a given strategy will have on a community. This means critically examining *how* adaptation occurs on the ground and the *impacts* of these processes, particularly in regard to the disproportionate burdens that adaptation imposes on the most marginalized members of society.

IN THE SHADOW OF THE SEAWALL

Seawalls are my entry point into the dilemma of placekeeping. Nearly one billion people live on low-lying coasts vulnerable to climate-related hazards.[31] As coastlines erode and waters rise to reclaim the world that humans have made, seawalls have become objects around which placekeeping is being negotiated, reimagined, and reconfigured through shifting logics of coastal protection and adaptation. On the shore, seawalls intersect with myriad desires in grounded and place-specific ways. They anchor reclaimed land in place, affix artificial islands in shallow lagoons, provide a promenade for recreation, and offer a feeling of safety from the next storm or tsunami. They also harm the environment, shift vulnerability downstream, lead to the disappearance of the beach, and create cycles of dependency and centralized development. They engage the nostalgia for the old shore and the anxiety about what the shore might become, solidifying momentary expectations of permanence and temporality.

Seawalls provide a snapshot of placekeeping and the ways in which desires for permanence collide with oppression, attachments to place, and anticipation of loss. Through seawalls, placekeeping comes to life, illuminating multiple dimensions of injustice and pathways of adaptation, inviting us to consider the larger sphere of cultural work that foregrounds place-based struggles for perseverance. Seawalls mark sites where disruption and regeneration exist side by side, allowing us to see placekeeping in different political lights.[32]

As infrastructural entanglements, seawalls are sites through which social, political, and environmental systems are imagined and materialized. The infrastructural turn in the social sciences centers on understanding complex entanglements of social and material life through the lens of infrastructure.[33] Ash Amin writes that material infrastructure, utilities, and services of the built environment are deeply involved in the making of place and the shaping of urban social identity.[34] This extends to the making of oppressed and colonial spaces.[35] Understanding infrastructure in this way requires us to reach beyond time and space to see the grounded ways in which we are bound to each other's pasts, presents, and futures.

Built to divide, to protect, and to contain, walls are central to the human project of settling in place. As such, they enter into some of the most contentious terrains of social life, where possession and belonging result in battles between individuals, communities, nations, and the natural world. Even in the landscape of the imagination, where fears and curiosities play out unhindered by the laws of physics, walls speak to the existential struggle of preservation. They are conjured to critique societal incongruities and contested definitions of place. Octavia Butler built them as a symbol of futility in her acclaimed novel *The Parable of the Sower*.[36] Ursula K. Le Guin demolished them alongside hierarchy, superiority, and greed in *The Dispossessed*.[37] Marlen Haushofer trapped her protagonist behind one at the end of the world in her aptly titled novel *The Wall*.[38]

Rachel Carson observed the colonization of seawalls as a "most fascinating problem," whose impact would seem to render life "doomed to futility" yet around which a distressed and resilient vitality has emerged.[39] Referring to the nonhuman struggle for perseverance, she writes, "For whenever a seawall is built . . . these hard surfaces immediately become peopled with typical animals of the rocks."[40] People too are animals of the rocks. Humans persevere with seawalls, bonded in mutual futility. They rely on them, and they look to them for protection and permanence amid the harsh realities of the natural world and its disruption. And similar to the fauna that inhabit the seawall, humans express what Rachel Carson observed as the pressure to persist, "the intense, blind, unconscious will to survive, to push on, to expand."[41]

Yet seawalls do not always result in the landscapes of which their builders dream. Not only do they embody politics that seek to control the natural world; they also invite the unexpected, including creative and informal avenues of maintenance. The notion of "resourcefulness" speaks to the limits of what people can do with the problems they have inherited from capitalist disruption.[42] People must make use of whatever means are within their reach to manage vulnerability. Alejandro De Coss-Corzo describes this phenomenon as a "patchwork" of repair in the case of Mexico's networked water systems, where workers maintain infrastructure in direct relation to "decaying budgets, precariously low wages, ongoing material ruination, the unpredictability of unwanted water flows, and a widespread lack of materials and tools."[43] Seawalls similarly call for a patchwork of interventions as they exist in perpetual decline.

Recently seawalls have become a subject of deep contestation. There is a growing divide between those who press for adaptation in the form of "hard" structures and those who advocate for newer and more experimental "soft" measures modeled on natural processes. Seawalls are seen as both a response to rising seas and a cause of erosion and climate vulnerability. For those who live behind seawalls, caught between the internal and external pressures of changing political and environmental realities, a line has been drawn and a shadow has been cast.

The shadow of the seawall, in fact, is an apt metaphor by which to understand the different sides of this existential dilemma. Imagine a shadow cast by a seawall. As Earth rotates around the sun, the shadow moves from one side to the other. The shadow first falls landward and traces an attachment to place and a nostalgia for the past. Under this shadow are the bitter traces of historical injustice and the roots of coastal vulnerability that haunt the landscape. As the sun crosses the zenith, the landward shadow disappears and a second shadow moves seaward, alluding to the anticipated loss of place. Under the seaward shadow lives the anxiety of disappearance, the fear and desperation of loss of place.

Somewhere between these two shadows is the struggle to maintain human life on the shore. In some cases, this translates into maintaining a seawall inherited from a colonial past, no matter how futile the effort. And in other cases, placekeeping entails preventing seawalls from carving out spaces of privilege in the name of climate adaptation. Either way, seawalls have become ideologically and physically entangled in the making and keeping of place, transcending their material form and entering into the ephemeral, enduring, and evolving interconnections of daily life.

A GLOBAL ETHNOGRAPHY OF PLACEKEEPING

Placekeeping is a process that evades conventional study, calling instead for a methodology that crosses disciplinary boundaries as well as boundaries of time and place. My approach emerges from an epistemology that situates the experiences of people who are often excluded from the knowledge-making process at the center of analysis. This entails moving away from conventional and positivistic traditions of social science research by engaging with Marxist, feminist, antiracist, Indigenous, and Third World perspectives that challenge the coded imperialisms of mainstream methodologies. While qualitative research as a whole has moved in this general direction, the meanings and intentions of research that calls itself critical varies. I understand critical research as a process that strives to be a "reflexive, empathetic, collective, self-altering, socially transformative, and embedded exercise in knowledge production."[44]

The complexity of coastal vulnerability also requires that we slow down, that we take time to pay attention to the particularities and interconnections of disrupted shorelines and embrace the challenge put forth by Michael Burawoy to comprehend the world as a set of fragmented, dispersed, and volatile lived experiences characterized by contradiction, struggle, and change.[45] This is the objective of global ethnography, a set of approaches that moves beyond traditional anthropological frameworks of focusing in depth on one place and involves looking instead at placemaking projects, where new kinds of places are being shaped by global forces.

While global ethnography provides a very powerful tool for understanding contemporary social problems, there are some limitations in its application to placekeeping that I have sought to rectify in my approach. Global ethnography, as it currently stands, seeks to understand processes of global destabilization in a way that tends to privilege abstract processes of globalization while overshadowing the very real and material political landscapes of place-based existence. Sea change and coastal defense entail destabilizing relations that extend beyond most operational definitions of globalization, sometimes by thousands of years. Furthermore, climate change is not isolated to the social forces that constitute globalization. Burawoy has expanded the concept to gauge epochal change through an extended case method that spans place and time through a combination of data gathered by means of archival research and participant observation. Yet global ethnography remains committed to explaining social transformation through unidirectional or prescriptive relationships that emerge within a global ethnographic framework; this takes the form of ascribing agency exclusively to external forces by explaining place-shaping processes as a flow that disrupts traditional social relations.

To resolve these limitations, I conceptualize global ethnography within the wider frame of the Anthropocene, a geologic era defined by human-damaged landscapes rather than natural fluctuations. Globalization plays a role in shaping the Anthropocene but is by no means the only set of forces involved in the destabilization of communities, ecologies, and social relations. I took inspiration from Anna Tsing's conceptualization of an *ethnography of global connection*, defined as a place-specific study of diverse and conflicting interactions that occur within a "zone of awkward engagement."[46] Tsing's approach differs from Burawoy's in one particularly important way: while Burawoy aims to generalize the connections between macro-level and micro-level social processes, Tsing aims to simply untangle landscapes embedded with specific and historical social connections.

To further complicate my own understanding of what global ethnography measures, I had to account for the fact that while people are central to the story of climate disruption, they are not the only players. This is why Jane Bennett challenges us to pay attention to the shifting temporalities and dynamic vitalities of the material world.[47] Anchoring seawalls as objects of investigation for a global ethnography opened new possibilities for gaining deeper insight into the social and material assemblages of vulnerability and practices of placekeeping. As I was conceptualizing my research on seawalls, I was careful to avoid working from the top down and the outside in. Instead, I focused my attention on seawalls as objects rooted in place and around which competing values are being negotiated across local and global fields of power and flows of knowledge.

Global ethnography works well in combination with a place-sensitive ethnographic perspective that is attuned to the social and material dynamics of infrastructural assemblages that shape and give value to processes of placemaking. These combined methodological approaches make it possible to address the complexities of human efforts to adapt to sea change, a problem that is predicated on historical inequality where flows and connections play out differently across place and time in unequal and diverse social landscapes. Together, these approaches allowed me to study both the material and cultural dimensions of placekeeping alongside the forces of global disruption. Bridging Tsing's place-specific model with Burawoy's global forces model enabled me to untangle two landscapes caught up in a common infrastructural dilemma.

The strength of this approach lies not so much in the ability to compare cases but rather in the ability to trace what is important to people

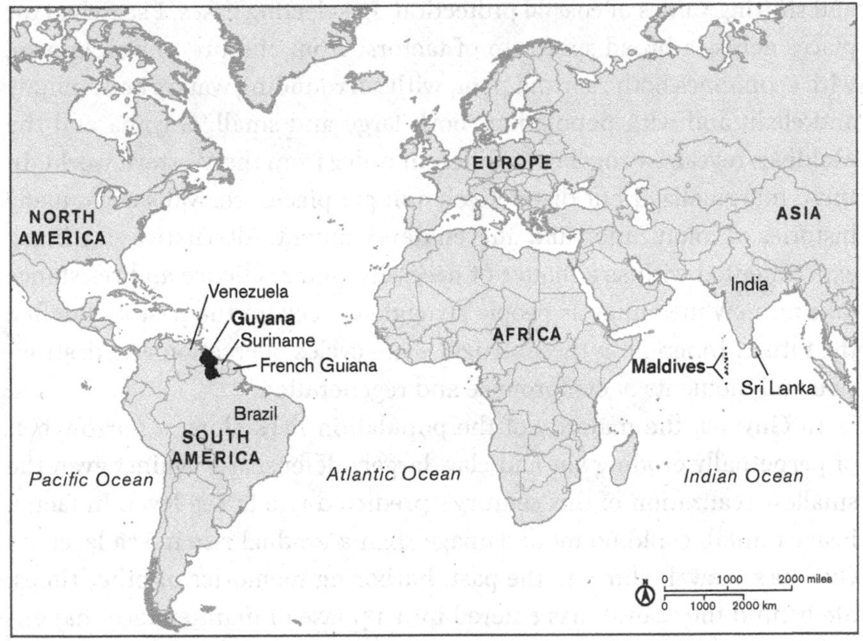

Map 3. Connected in a common struggle: Guyana and the Maldives

and what informs their desire to shape their environments across multiple contexts. The juxtaposition of cases sheds light on the material and cultural factors that enable or prevent placekeeping in contexts of disruption. Because of the extreme limitation of resources worldwide and the impossibility of fortifying all the world's populated coastlines, priorities rise to the surface that will determine the future map of the world. Studying placekeeping from this perspective is thus one attempt to gauge the bigger picture through the lived experiences of people interdependent with their coastal environments.

THE CASES OF GUYANA AND THE MALDIVES

Both the Maldives and Guyana are frontline communities entangled in seawalls. Both are undergoing placekeeping struggles under conditions of climate disruption while negotiating with competing logics of adaptation and shifting values of coastal protection. In selecting cases, I searched for places across a broad spectrum of factors: from obscure to well known, with economies both rich and poor, with surrounding waters both muddy and clear, and with populations both large and small. Guyana and the Maldives reveal a struggle of a different order from the Western world. In these marginalized but distinctively unique places, seawalls accompany histories of colonization and uneven development. Alternatives are being experimented with as a matter of necessity, and resilience and resistance take on new meanings as people struggle to reclaim the past and define the future. Inherent in these struggles are cycles of creation and destruction and moments of compromise and regeneration.

In Guyana, the majority of the population lives along a narrow belt of perpetually eroding silt and clay, largely defenseless against even the smallest realization of this century's predicted rise in sea level. In fact, a heavy rainfall could do more damage than a gradual rise in sea level. As Guyana's seawall clings to the past, harboring memories of other times, life behind the seawall has entered into a phase of maintenance that entails an uncomfortable negotiation with the colonial landscape and its accompanying burdens. By its very nature, Guyana's seawall is an archive of struggle, an aggregate of debris that warps, rusts, erodes, and decays against the salty breath of the ocean.

The geography of Guyana lends to its unique problems as well as its incredible vitality. Beyond the coast of Guyana, mountains tower high above an immense rainforest. To reach them requires an overnight journey up the Essequibo River or, for those who can afford it, a ride in a small prop plane. From the clouds, the hinterland dissolves into a sea of green, giving the impression of permanence. It is easy to take for granted the ghostly process of geologic time and miss the rhythm of the moving parts below. Over millions of years, a bed of sand was carved by rivers that gnawed at the mountains and washed the pieces to shore. Together, these waters expanded the continent, soaking its outer edge in a wilderness of mangroves and mud, giving meaning to *Guyana*, an Amerindian word that translates as "land of many waters." The Guyanese poet Wilson Harris once described the unfolding terrain as "a great magical web born of the music of the elements," its many rivers "the arteries of God's spider."[48]

Today, the alluvial waves of the Atlantic Ocean push against a seemingly endless seawall. Fragments of granite and concrete push back at the tides, protecting an artificial basin that sinks six feet below sea level, stretching inland for five to eight miles. Two hundred miles of the coastline are engineered this way, drained and reclaimed over centuries for the cultivation of sugarcane. The muddy embankment on the landward side of the seawall is sliced by perpendicular canals of varying sizes, the largest of which end in guillotine doors that lift to release water before falling to cut off the sea. The political frictions of contemporary Guyana, from colonialism to democracy, are multidimensional, involving the right to build a new world out of the fragments of an old one. Here seawalls are already so entangled in place that their abandonment would inflict further harm on populations already subject to historical injustice.

On the other side of the world, in the small island nation of the Maldives, another seawall entanglement tells of the struggle to adapt to sea change. Located in the Western Indian Ocean, the Maldives is the lowest-lying country in the world. The highest peak is a mere 2.4 meters. With 1,192 coral islands arrayed in a double chain of 26 atolls, the Maldives appears as a shimmering vision of cosmic wonder, "scattered like the Milky Way over fifty thousand square miles of sea."[49] The islands themselves are dynamic ecosystems that dance in slow motion to the rhythms of the wind, sand, and sea—each immersed in cycles of creation and destruction like a thousand sand mandalas.

By many accounts, the Maldives is the poster child of anticipated loss. Images of knee-deep islanders in crystal blue waters are conjured in books, journals, works of art, films, and other media captioned by the term "climate change refugees." But what is happening in the Maldives is far messier, and the heart of the matter is drowned out in the sensationalized narratives of well-intended Westerners and Maldivian politicians alike. As most stories wash over the political history of development, infrastructure, and political repression in favor of evocative narratives intended for Western audiences, they miss what is really at stake in the Maldives: a fight for the future. This fight, which is deeply connected to practices of placekeeping, is based in a long history of political greenwashing and infrastructural development.

For some, seawalls are the heroes of the Maldives, emblematic of so-called resilience and adaptation. Yet it is easy to forget that these structures were not always regarded as such. Seawalls entered the story long before adaptation came to dominate development priorities and discourses in the Maldives and throughout the world. Ushered in during a period of rapid development, with the help of Japanese engineers, modern seawalls in the Maldives find their origins in the makings of modernization, accompanied by a political struggle to define place. Now, as the sea changes, the anticipated loss of place has opened the playing field, giving rise to divergent visions of the future and new ideas for keeping place where it is. Sensationalized narratives of climate change—including tropes of victimization and heroism—are utilized by both ends of the political spectrum, often serving to legitimate lesser-known environmental injustices through the proliferation of projects that displace people and damage the natural ecosystem's ability to adapt to sea change.

OUTLINE OF CHAPTERS

The chapters that follow trace the intricacies of placekeeping in Guyana and the Maldives, illustrating how seawalls are entangled in histories of oppression that inform attachments to place and anxieties of loss. Each case is treated separately and organized in paired chapters to emphasize placekeeping as a process enacted through efforts to *make* and *keep* place. Returning to the metaphor of the shadow, each pairing accentuates the different sides of the dilemma of placekeeping—from the perspective of a landward shadow that draws attention to the past and the making of coastal vulnerability to that of the seaward shadow that alludes to the anticipated loss of place and efforts to stay in place.

Chapter 1 contextualizes placekeeping on the shore through a global history of coastal protection that Guyana and the Maldives share. In contrast to the limited future-oriented framework of adaptation, I emphasize the historical injustices of coastal vulnerability through overlapping waves of colonization, modern development, and globalization—the foundation on which contemporary placekeeping is negotiated. The danger of rising sea levels is not a novel problem; people have confronted change along the coast in various ways for centuries. I describe how efforts to stay in place engage with complex entanglements of infrastructural systems, politics, and shifting values of coastal protection. In particular, desires for conventional gray infrastructure are increasingly coming into conflict with emerging desires for nature-based solutions and green infrastructure.

Chapter 2 turns to the story of Guyana's two-hundred-mile stone and concrete seawall, which is embedded in a landscape of political, economic, and racial injustice. I trace a two-hundred-year history of sea defense and land reclamation, from Dutch colonial reclamation to the end of the Burnham dictatorship in the mid-1980s, to show how seawalls are used to *take* place. Through this history, we are offered a glimpse into the formation of coastal vulnerability as one of the many lingering impacts of colonization. Guyana's seawall maintains an indelible reference to the colonial past. Therefore, through shifting politics, from slavery to independence, the seawall remains as both an opportunity and a burden.

Chapter 3 continues to unpack Guyana's placekeeping struggle, exploring efforts to negotiate with shifting values of coastal protection. It focuses on efforts surrounding the planting of a "green seawall," exploring the ordinary and extraordinary ways that placekeeping efforts are utilized to maintain the aging seawall in contexts of disrepair, impoverishment, rising tides, and political corruption. Placekeeping is being negotiated alongside a complicated sense of belonging that Guyanese feel in relation to the fertile coast and its troubled past and highlight the linkages between engineered resilience and colonial nostalgia. Redefining coastal protection means negotiating with the past and attachments to place.

Chapter 4 journeys to the other side of the world, to the islands of the Maldives. The profound impact of anticipation—of sea level rise and loss of place—is portrayed through the story of the Maldives, which has been transformed through land reclamation schemes and seawall construction over the past fifty years. Like Guyana, seawalls in the Maldives embody histories that are place-specific, but they also become politicized through anxieties about the future and anticipations of loss. Seawalls are understood through competing logics of adaptation constructed in a landscape of competing desires, from authoritarianism and neoliberalism to a movement for democracy. Surrounding the drama is a seawall that was built by the Gayoom dictatorship and concretized by the Japanese government in the mid-1980s. Adaptation discourse assumes that land scarcity is a fixed condition of ecology, ignoring how land is a resource that is unjustly made and unjustly distributed through urbanization and development. In the Maldives, seawalls are part of this unjust allocation through their placement and planning.

Chapter 5 delves deeper into the dilemma of placekeeping in the Maldives and explores how contested visions of the future gravitate around seawalls and their alternatives. Over the past decade, the Maldives has become a place of intense anticipation and a battleground for climate adaptation. With the anticipation of runaway climate change, environmental narratives—including the spectacle of sinking islands—are increasingly shaping national development priorities in low-lying countries. However, without a firm commitment to small island justice, adaptation threatens to further exacerbate uneven development, calling into question whether many of the smaller ancestral islands should have a place in the country's plans for survival. Three contrasting efforts to define what is worth saving are explored: the authoritarian government's controversial population consolidation scheme to create fortified "safe islands," a plan to create a Maldives where islands are regenerated and connected through systems of transit and mobility, and a partnership between international designers and a youth movement to create artificial reefs and living seawalls.

I end by turning to the question of what seawalls tell us more generally about future placekeeping efforts around the world. While our collective future of seawalls is likely a dark one, there are many possibilities for embracing a more critical approach to adaptation; in return, we are shown a pathway to climate justice. However, if we ignore the social complexities and hidden struggles of placekeeping and the logics behind these practices, the cycle is bound to be repeated. Without attention to people and their relationships to seawalls and without a clear understanding of coastal vulnerability as a process that engages with human interventions and systems of inequity, it is unlikely that visions of the future will reflect the reality of coastal transformation, even in their most pessimistic forms. We must not lose sight of the infinitely embedded social and spatial contexts of coastal disruption that make solutions unscalable and staying in place a phenomenon of necessity. These are the scales at which the contours of sea change are shaped, where humans strive to make the best possible world out of the uneven playing field they have been handed.

1 Coastal Disruption

Shortly after Superstorm Sandy made landfall, I found myself passing through New York City on my way to Guyana, my mind filled with the surreal images that dominated the national media of waves crashing over seawalls, water rushing into New York subways, trees bending backward, and lights flickering out over flooded city streets. Sandy's destruction was still unfolding as I passed through JFK International, and the bits of conversation I picked up on my way through the airport were heavy with anxiety. The scope of the local destruction was shocking enough. Staten Island, the New York borough barely twenty miles from JFK, had been inundated, with businesses, homes, and lives lost. But perhaps equally troubling was the growing realization, among local residents and national onlookers alike, that Sandy was not an aberration, a once in a generation event that would fade into memory after the debris was cleared. As I made my way to the plane, a security guard told me that this is going to happen again and more often.

The slow-moving emergency of sea change makes it nearly impossible to visualize the future across the broad and uneven terrain of coastal existence. Sea change is not an entirely exogenous force like the wave of a thousand-year tsunami. It is both a slow process and an immediate one that evades traditional modes of representation, forcing us to rely on maps that employ a combination of science and imagination to gauge the many dangers of our changing climate. These maps of the future are everywhere, dominating the epistemological landscape of sea change, from the sensationalized and glossy depiction of a hypothetical worst-case scenario to the mundane forecast of a localized flood map rendered to account for predicted sea level rise. One of the most widely disseminated maps was published in 2013, in *National Geographic*'s dramatic cover story on rising sea levels.[1] The fold-out map vividly illustrates how Earth's continents would be reshaped if all the ice in Greenland were to suddenly melt. Flipping through the pages, the image compounds into a horror show worthy of a Hollywood eco-disaster film and eliciting all the emotions one feels when watching the world fall apart. The scientific picture is no less dramatic.

Such images share a likeness to Alan Weisman's depictions of the built environment in *The World Without Us*, in which he imagines the fate of our modern infrastructural systems devoid of humans to maintain them.[2] With people largely absent from these forecasts, the central protagonists are the structures and the imaginary lines that delineate familiar streets and borders. While these evocative images have the power to change how we see ourselves and the world around us, it is easy to forget that these structures engage with the rhythm of human politics, human values, human agency, and human oppression. They are also the products of unanticipated consequences and spur-of-the-moment decisions. Their perseverance or decline is not inevitable. Nothing is ever as simple as it seems. The problem is that as analytically accurate as our predictions may be, they cannot account for the people—their worlds and their lives—through whom they play out.[3]

To make sense of the present moment and our worst fears about the future, including the unique struggles of Guyana and the Maldives, we must first pay attention to the past. The sea as we know it stabilized six thousand years ago—long enough for empires to colonize Indigenous lands, modern cities to form, extractive industries to plunder Earth's natural resources, and fossil fuel companies to destabilize the climate. Over thousands of years, waves of disruption have pushed, squeezed, and eroded life on the shore, and within these waves, we find conflicting notions of progress emerging from projects of colonization, development, globalization, and, more recently, adaptation.

The battle to hold back the sea spans centuries. Early seawall formations accompanied the rise of global capitalism, the expansion of colonial rule, projects of nation building, the intensification of agriculture, industrial production, tourism, and the growing importance of infrastructure for geopolitical power. More recent seawall formations have branched into new design paradigms that seek to offer greener and more aesthetically pleasing alternatives, mimicking the natural world while blending into the built environment. As coasts are made and remade, they are negotiated across shifting political and environmental values. In this way, seawalls are political.

Langdon Winner, who helped define the field of science, technology, and society, once asked if infrastructural artifacts have politics.[4] He explained how in some cases technologies may embody *deliberate politics*, citing the example of low-hanging highway overpasses around New York City.[5] The overpasses were designed to prevent people of color—who relied on buses and public transportation—from having access to public beaches. However, many modern technological systems do not necessarily contain deliberate politics. Rather, there are other and more complex ways in which technologies can embody politics. For example, when man-made systems require, or are strongly compatible with, particular kinds of political relationships. Winner calls these *inherent politics* and draws on the example of nuclear power as necessitating social hierarchies and authoritarian forms of governance.

As large-scale infrastructural systems, seawalls embody inherent politics that call into play hierarchies of power and knowledge. They also accompany processes of uneven development and uneven protection. Used to discipline the shore, seawalls emerge from capitalist desires for urban expansion, economic growth, development, and protection. According to Andreas Malm, "seawall politics" are entangled in the contradictions of capitalism, often used to protect capital investments rather than people.[6] Yet seawalls are not only drivers of capitalism; they are also embodiments of ideas and systems of value that are communicated, desired, and socially constructed. Seawalls are what Susan Leigh Star and James R. Griesemer call *boundary objects*.[7] Boundary objects are recognizable across time and place but malleable enough to adjust to different social contexts and local conditions. Isto Huvila proposes that we consider all built spaces and infrastructural projects as boundary objects so as not to forget that physical structures, like words, have shifting meanings.[8]

This chapter offers a political history of seawalls and coastal disruption across multiple and overlapping waves of injustice to show how pathways of adaptation are being negotiated through complex entanglements of power and competing logics of protection. Placekeeping is shaped by systems of oppression and structures of global inequality instituted during eras of colonization, modern development, and globalization. With the gathering momentum of ocean instability wrought by centuries of industrial-scale human intervention, and after the thousands of years of coastal modification—starting with the labor-intensive processes of digging, pumping, and dredging and the patchwork of attempts at coastal stabilization—coastal vulnerability has come to emblematize the gross inequalities and injustices of contemporary climate change. Now, as seawalls come under scrutiny, they lie at the very foundation of the modern problem.

WAVES OF INJUSTICE

Coastal vulnerability as it is experienced today is one of the many manifestations of the modern project, a political and philosophical movement that emerged in Europe during the seventeenth- and eighteenth-century period of the Enlightenment. At the core of the modern project is a penchant for social and physical order aided by science, a tool for imagining, legitimizing, and expanding modern rationality and social order to far-reaching and often extraordinary domains. This includes the manufacturing of new frontiers for the exercise of power, the accumulation of capital, and, more recently, the manipulation of life itself. These desires stem from a frustration with chaos and a utopian idea that the eradication of chaos would necessarily lead to a better world.[9] As such, the modern project is host to a range of competing desires: the unbridled pillage of environmental resources, the promise of sustainability, the amelioration of poverty, and the pursuit of profit.

The utopian dream of modernity has become a nightmare for many, presenting a paradox of cascading emergencies. Karl Marx, Joseph Schumpeter, and other critics of the post-Enlightenment social order describe modernity as a process of creative destruction that ushers the world into a state of permanent emergency and endless adaptation.[10] As Theodor Adorno and Max Horkheimer somberly stated over half a century ago, "The fully enlightened world radiates disaster triumphant."[11] Marxian and Schumpeterian notions of creative destruction speak to the internal contradictions of capitalism that produce instability and crisis as a central tenet of capitalist development and expansion. The resulting relationship is paradoxical; capitalism thrives on its own destabilizing tendencies and profits from the corrective measures that it necessitates, including adaptation and resilience building.

In a process of perpetual creation and destruction, modern expansion creates the conditions for its own demise while offering up its own remedies in the form of sustainable development and other green, regenerative measures that aim to perpetuate economic growth. Modernity entails the formation and exhaustion of frontiers and the never-ending demand for breakthroughs. Out of this cycle come damaging innovations and false solutions. As the cycle repeats itself, shortsighted development leads to maladaptation, and maladaptation drives the development of capitalist innovations designed to restabilize and maintain the status quo.[12] This cycle is further amplified during times of crisis. As the journalist and political activist Naomi Klein argues, elites are now more blatantly capitalizing on disaster to pass draconian economic policies that in less distractingly urgent times would have generated public scorn—a phenomenon Klein names *disaster capitalism*.[13]

Overlapping waves of colonization, modern development, and globalization constitute the modern project. These are the drivers of global inequality and the progenitors of coastal vulnerability—the foundation on which contemporary placekeeping occurs. The first wave—colonization—entailed the dispossession of Indigenous lands and many of the world's coasts for access, extraction, and settlement by European colonizers, particularly by those of the Dutch, French, and British Empires. Justified by notions of social progress, new boundaries were drawn, ancient territories were violently parceled, and the colonies were physically shaped for the purposes of empire.[14]

During the period of European colonialism, unimaginable violence swept the planet. The scale was so profound that it changed the climate. One can observe this by looking at Antarctic ice core records between 1570 and 1620, a period that Simon Lewis and Mark Maslin call "the meeting of the Old World and New World."[15] During this time, a sharp decline in atmospheric carbon dioxide can be traced to a rapid decline in human populations from disease, famine, and genocide, leading to the regeneration of forests and grasslands for carbon sequestration. The lowest point occurred in 1610, known as the "Orbis spike."[16]

As the reach of empire extended across oceans, it brought with it practices that set into motion cycles of construction and abandonment, enabled by slave and prison labor, with little constraint or concern for human and non-human life. "Free enterprise ruled the day," one British historian noted, "and there was little real concern for the environment."[17] Not only were colonial lands stripped of resources, but they were also converted into recreational enclaves and treated as the laboratories of modern knowledge production and experimentation. In Australia, colonized coastal land was claimed by the British Empire explicitly for recreation and relaxation.[18] In the Chao Phraya Delta of Thailand, British and Dutch experts adapted land reclamation and terrestrialization practices in the colonies, expanding their knowledge for the benefit of empire.[19] Similarly, in Guyana, colonization converted the coastline into an agrarian economy, and the first seawalls emerged to protect colonial assets from inundation.

The second wave of injustice—the development project—emerged in the mid-twentieth century with the creation of the United Nations (UN), conceived in 1945 as a body for promoting peace and prosperity after World War II. Development, understood in economic terms, entails an effort to sustain high national economic growth rates and to transform agrarian economies into industrialized systems of production. Definitions of development vary according to a range of political and philosophical assumptions that are often contentious. Wolfgang Sachs, a prominent and critical scholar of development studies, describes development as a concept of "monumental emptiness," encompassing an array of conflicting perspectives and notions that embrace both the enhancement of economic growth and the desire to help the poor and powerless.[20]

Since its institutional inception, development policy has been fundamentally concerned with the economic trajectory of nation-states, arguably part of a longer tradition of management and planning carried over from nineteenth-century processes of city planning and nation-state building in Europe.[21] Following the 1944 Bretton Woods conference, the UN decreed that its global objective would become the promotion of a "rising standard of living," to be measured by per capita gross domestic product (GDP).[22] In this commitment, and with the World Bank's and the International Monetary Fund's objective of integrating developing economies into the global economy and reducing poverty by promoting economic growth, the UN became "the mother church of developmental optimism."[23] Around this time, UN member nations signed the General Agreement on Tariffs and Trade, institutionalizing the organization of an international economy. Established in 1961, the Organization for Economic Co-operation and Development was created to design and implement development policy while promoting consensus among governments and the private sector.

During this time, attention shifted from European reconstruction to the project of developing former colonies, now liberated nation-states. However, colonization did not end with the onset of development; it took on new forms and intensified.[24] Uma Kothari, a scholar of postcolonial studies, traces the linkages by looking at the narratives of individuals who were involved in both the administration of colonial power and the project of development.[25] In one interview, a former colonial officer revealed how closely linked the two projects are: "If I can no longer do this job and work out here the next best thing is to be working for the development of Kenya in the development field—after all, it is the same thing."[26] Colonial officers and the early experts of development studies shared a common set of specialized skills that required knowledge of geographically isolated areas and local languages. As such, development should be understood as a rebranding of colonization.

The logic of development was driven by theories of modernization, which furnished the nascent project with a scholarly identity and a sense of scientific rationality and legitimacy. Drawing on the United States as the archetype of economic and social development, modernization theorists sought to identify and explain the conditions that have prevented underdeveloped societies from industrializing.[27] Concerned primarily with mapping the transition from "traditional" to "modern," modernization theorists conceptualized social transformation as a process of political, economic, and cultural differentiation.[28] In these early studies, the presence of modern Western cultural and psychological traits was viewed as "necessary to liberate Third World economies from traditionalistic restrictions to rational dynamism."[29] As such, tradition was considered a negative or impeding force in the process of development.[30] Furthermore, steady economic development became viewed as a necessary precondition for political democracy and the dissolution of radical working classes in developing countries.[31] Threatened by the increasing power of the Soviet Union, the modernization perspective viewed communism as socially, politically, and economically backward.[32]

The third wave of injustice—globalization—emerged at the turn of the twenty-first century, as development policies fell into crisis. A shift in international development discourse from "economic nationalism" to "world market participation" constitutes what Philip McMichael calls the shift from the development project to the globalization project. The globalization project, according to McMichael, "succeeded the development project—not because development is extinct, but because its coordinates have changed."[33] The basic definition of globalization holds that recent advances in transportation and communications technologies have led to an unprecedented degree of global economic and political interdependency and that new networks have emerged through the changing flows of capital, information, ideas, people, and resources.[34]

The shift from managed national growth to world market participation involved a massive restructuring of national economies in the developing world made possible by the rise of a global banking system and cross-border financial liberalization policies known as neoliberalism. Economic restructuring was presented as a solution to the crisis of poverty, further compelling developing countries to look outward rather than inward for their development stimulus. This included a dramatic reduction in public spending, the devaluation of domestic currencies to inflate the prices of imports and reduce the prices of exports, the intensification of export production, the privatization of state enterprises, and the deregulation of labor.[35]

The appeal of neoliberalism worked against the fundamental tenets of conventional development practice, placing the levers of economic growth in the invisible hand of the market rather than the iron fist of the state. Massive and unregulated private bank lending to developing countries followed, leading to the debt crisis of the 1980s, further debt traps, austerity measures, and other forms of neocolonial dispossession. The amplification of neoliberal policies has rendered affected communities increasingly vulnerable to even small disasters. In the case of Puerto Rico, for example, the legacy of US colonialism and the island's current status as a US protectorate have left the island's government in a cycle of dependency and debt.[36] In the Maldives, globalization has enabled the funding of seawalls and other large maladaptive infrastructural projects—a trend that continues to take shape in the growing awareness of sea level rise.

INFRASTRUCTURING THE MODERN SHORELINE

Within the projects of colonization, development, and globalization are the makings of the physical and social landscapes now endangered by climate change and the foundations on which contemporary struggles for placekeeping exist. Well before anthropogenic sea change emerged as a known problem, moving shorelines challenged modern desires for control, inviting efforts to counteract uncertainty, change, and instability through human intervention. Efforts to stay in place now involve complex entanglements of infrastructural systems, including technologies of coastal modification, that are built in relation to the drivers of global inequality.

The discipline of coastal engineering emerged in relation to the projects of colonization, development, and globalization. As such, coastal engineering mirrors the contradictions of capitalism, creating its own crises and need for remedies. The goal of coastal engineering is coastal protection, which is broadly defined as the modification of the shore for the purpose of maintaining or "improving" human uses of the intertidal zone. This can include the construction of seawalls, the reclamation of land from the sea, and the building of artificial islands, sand dunes, and other landscape modifications (including restoration projects) intended to further protect the human-inhabited coast.

The textbook history of coastal engineering celebrates a heritage of thousands of years of port development, reclamation of land from the sea, and coastal hazard protection, overlooking the social and cultural circumstances that brought these systems into being. In conventional accounts of the discipline, coastal engineering is presented as one of the many heroes of the modern world. Prior to the development of coastal engineering science, the relationship between humans and the shore was one of perpetual struggle and retreat. Ancient and premodern civilizations made use of natural harbors formed in the aftermath of glacial melting for protection. These early harbors were selected for their natural protection and ease of access, but as coastal dynamics changed—which they often did, and dramatically so—they were abandoned. Vikings, for example, relied heavily on natural harbors and would sometimes reinforce bays by sinking ships, but they could not compete with the natural closing of the Limfjord and the disappearance of natural harbors on the North Sea coast.[37]

Over time, as the sea stabilized and modern science emerged, retreat was no longer inevitable. Early techniques of coastal modification gave birth to new and more ambitious plans when coupled with the newly forming science of coastal engineering. In the early formation of Germany and the Netherlands, the first traces of modern engineering were born, and after millennia of retreat, the approach transitioned from retreat to "accommodation."[38] It is also no secret that as humans got better at coastal engineering, new, often disastrous, problems emerged. Early applications, poorly managed and piecemeal in nature, generated a false sense of security built on a foundation of environmental disruption.[39]

Prior to World War II, the disciplining of the world's coasts, including the colonization of coastlands for resource extraction and mining, paved the way for the modern capitalist system, with its flows of commodities, violence, and disease. Colonial technologies of coastal stabilization were used to expand access to a wide variety of resources through the reclamation of land from the sea and the fortification of coastlines. Coastal disciplining for the sake of ownership and resource extraction was largely responsible for initiating the racialized inequities of climate change, sowing the seeds for a lifetime of coastal vulnerability. For this reason, Ann Laura Stoler, in her edited volume, *Imperial Debris: On Ruins and Ruination*, calls on scholars to "track the 'concrete trajectory' of colonial exclusions and derailments that carve out the structures of privilege, profit, and destruction today."[40]

The creation of harbors opened access to remote corners of the world, and land reclamation helped expand territories for extraction. Early modern ports of the seventeenth and eighteenth centuries had become "major gateways for products, people and ideas that were exchanged throughout the world," which included resources extracted from newly claimed colonies.[41] As the centers of trade solidified, cities rapidly developed along the coast. The materials for industrialization came from the water, the ground, and the forest. Water was needed to power new machines, harbors were needed to land merchant ships, and rivers were needed for inland transport. As ships grew in size, larger harbors were needed, as were deeper canals and waterways for transportation.[42] This led to the development of a wide range of coastal infrastructural projects to accommodate modern industrial society.

The infrastructure of colonial expansion also set a baseline for the future of coastal settlement and protection. In New Orleans, for example, French levees were introduced to the landscape during colonization, setting a foundation for coastal vulnerability for years to come. Chandra Mukerji's study of French military engineering techniques during the seventeenth century argues that the failure of the seawalls and levees in the Mississippi Delta are a result of the colonial government's lack of commitment to sustain the stewardship necessary to maintain such an intricate system of works.[43] Stewardship was a "sham," she writes. "No one really cared for the city as a place that needed to be nurtured continually . . . as a precious and fragile physical environment, a gift of God that had been improved with human hands."[44] As Brian Larkin noted in the case of postcolonial Nigeria, colonial infrastructure has a lingering effect: "These intentions do not easily go away, and long after haunt technologies by shaping the conceptual horizons of what people expect them to be."[45]

After World War II, as global capital proliferated, so too did international development and loans to further fortify harbors for commerce and trade. Development projects ushered into existence a host of political and economic projects that have functioned to manage land, people, and natural resources. Many of these projects require proximity to the ocean, including nuclear reactor power plants (and their accompanying breakwaters, seawalls, docks, etc.), water treatment plants, and irrigation systems.[46] Rapid development has translated into the disappearance of beaches and in more extreme cases, the sinking of entire coastal cities.[47] In some cases, this was due to the introduction of hard structures; in other cases, the sand itself was being uprooted as an unregulated commodity.[48]

In 1950, the formal discipline of coastal engineering emerged as a distinct area of civil engineering, institutionalized with the first International Conference on Coastal Engineering held in Long Beach, California. In the volume *History and Heritage of Coastal Engineering*, edited by Nicholas Kraus, representatives from fifteen countries that have hosted the conference since its inception detail the history and development of coastal engineering in their respective nations.[49] In the preface, Kraus notes that "despite cultural differences, common themes are found" and notes three progressive cycles of coastal engagement: "(1) exploitation and utilization of the coast; (2) development of protection from coastal hazards such as flooding and erosion; and (3) preserving and creating harmony between nature and coastal uses."[50] According to Kraus, only the most advanced nations had entered into the third cycle of coastal engineering.

During this time, coastal defense mirrored the assumptions and aspirations of modernization theory. For example, Johan van Veen, a prominent coastal engineer from the Netherlands, believed that the Dutch held the secrets "to turn foreign waste lands into well-kept paradises"; they simply needed "the moral and effective support of the country in which they are working."[51] One passage in particular from Van Veen's account of Dutch engineering practices emphasizes this school of thought in striking detail:

> Dredging, draining and reclaiming might rightly be called the hope of the poor countries, because such positive works yield riches. Anyone who has traveled in the backward countries and marveled at the low standard of living there, has perceived a lack of good roads, good drainage, good rivers, good harbours, good houses, good crops, and much else. In those countries former generations did not build enough, they did not put enough into the soil. The people of those countries may work hard to keep themselves from starving, they may fight their neighbours to the very limit of their energy, but riches come from well-organized and intelligent effort to improve the land and exploit its possibilities. Fortunate is the people whose predecessors developed the country.[52]

As the discipline of coastal engineering crystallized, a formal platform emerged for the global exchange of ideas and information. Combining knowledge from the fields of oceanography, geology, and geomorphology, coastal engineering became employed to meet the demands of global capitalism. While coastal engineering maintains its place-specific contexts and cultural nuances, mutual efforts congealed around challenges connected to the international growth of domestic and global maritime trade and the commodification of the coast. Coastal engineering challenges emerged from desires to manage coastal hazards and the natural destabilization of harbors formed after deglaciation in order to secure coastlines associated with tourism, recreational beach use, agriculture, the expansion of fishing harbors, and industrial infrastructure, including nuclear power plants and offshore oil drilling.[53]

Coastal engineering played a seminal role in enabling the development of seaside destinations and residential coastal areas. While recreational coastal defense emerged long before the mid-twentieth century, the introduction of cars and artificial beaches marked a turning point in the history of coastal engineering during the era of development.[54] For example, in Australia, the emergence of private automobiles, coupled with the improvement of roads, "greatly accelerated the use of beaches, coastal lakes and estuaries as holiday and recreation areas."[55] In the 1950s, as France became a tourist hub, the government commissioned the development of artificial beaches to meet demand, including the construction of a completely artificial beach in Monaco between 1965 and 1967.[56] In Mexico, international tourism in conjunction with the national program "Marcha hacia el Mar" (March toward the Sea) to decentralize industry and the population necessitated coastal engineering for the accommodation of large cruise ships and ferries.[57]

Coastal engineering also played an important role in enabling oil exploration and offshore drilling as the demand for fossil fuels increased. Following the world oil crisis of 1973, oil drilling took to the deep sea, and wave science became a central focus of coastal engineering, explicitly tied to the construction of offshore platforms for oil and gas production.[58] In Canada, following the capsizing of the *Ocean Ranger*, coastal engineers shifted their focus to the creation of offshore structures for oil exploration and extraction, including artificial drilling islands. These early artificial islands captured the imagination and set a precedent for future projects of nation building as well as engineered pathways of adaptation for small island nations.

Throughout the mid-twentieth century, coastal armoring started to shift strongly to concrete, and coastal defense began to look the same across the many human-altered shorelines of the world. Previously, the stabilization of shorelines utilized available natural resources collected or mined from quarries. By and large, access to quarried rock was critical for marine projects of national importance.[59] The sheer amount of rock needed for marine projects is astounding; the reclamation and construction of the new Hong Kong Airport, for example, required 9.3 million tons of quarried rock per year over the course of five years.[60] Market-based concrete armor units emerged as a competitive alternative to quarried rock for large-scale projects designed to withstand the massive forces of wave energy.[61] Typically constructed from steel or glass-reinforced plastic molds, concrete units are often assembled on site in a large yard or area for casting and can be shaped in ways that increase stability.[62] Concrete is now among the most common materials used for coastal armoring, including the construction of seawalls.

THE ENVIRONMENTAL TURN

The environmental impacts of colonization, development, and globalization have been devastating, introducing a vast and incalculable array of environmental injustices. Deforestation, desertification, pollution, and biodiversity loss, combined with unprecedented levels of ocean acidification and stratospheric ozone depletion, constitute an ecological crisis of global proportions. Environmental movements throughout the Global South have formed in response to the devastation.[63] Two infamous examples of ecological disruption are the Bhopal disaster of 1984, which exposed over five hundred thousand people to methyl isocyanate gas leaked from a pesticide plant owned by Union Carbide, and the "Rainforest Chernobyl" of Ecuador between 1964 and 1990, when Chevron-Texaco extracted massive amounts of oil with obsolete methods, making no effort to protect the neighboring villages from the toxic runoff.[64]

Ecological decline has also posed new challenges for the existing frontiers of capital accumulation and expansion. James O'Conner, a sociologist and political economist, identifies this problem as the "second contradiction of capitalism"—that is, as capitalism expands, it inevitably runs up against the economic limits of its undermined conditions of production wrought by environmental degradation, leading to an eventual collapse.[65] O'Conner's perspective, while seemingly bleak, is optimistic; he posits that the revolution unrealized in the wake of the first contradiction of capitalism, as identified by Marx and Engels, will become realized in the wake of global environmental disruption. Yet John Bellamy Foster, another prominent sociologist and Marxist theorist, is not convinced and cautions that "we should not underestimate capitalism's capacity to accumulate in the midst of the most blatant ecological destruction, to profit from environmental degradation, and to continue to destroy the earth to the point of no return—both for human society and for most of the world's living species."[66] Wolfgang Sachs further warns of an emerging "global ecocracy" in which "capital, bureaucracy and science—the venerable trinity of Western modernization—declare themselves indispensable in the new crisis and promise to prevent the worst through better engineering."[67]

In the mid-1980s, "sustainable development" emerged as the dominant answer to the mounting problem of ecological disruption. Addressing the problem of resource depletion and the challenges it posed to economic growth, development was recast as a friend of the environment. Sustainable development, defined as development that "meets the needs of the present without compromising the ability of future generations to meet their own needs," was offered as the solution.[68] The environment became a central issue in international development policy with the creation of the UN World Commission on Environment and Development and the 1987 publication of *Our Common Future*. A growing recognition of the pervasiveness of environmental problems linked to economic activity resulted in attempts to create global regulatory mechanisms, including the 1989–92 Montreal Protocol on Substances That Deplete Ozone Layer, the 1992 Convention on Biological Diversity, the 1992 Framework Convention on Climate Change, and the 1997 Kyoto Protocol on Climate Change.[69]

However, as Gustavo Esteva, a Mexican activist and public intellectual, argues, the project of sustainable development merely entails a rebranding of development and has done nothing to question development itself.[70] Rather, sustainable development was explicitly conceived of as a strategy for *sustaining* development rather than supporting a return to an enduring and diverse natural and social life. For decades, scholars from Latin America and the Global South have recognized the contradictions of development, particularly that "modern" nation-states were able to attain that status only by exploiting the resources and labor of the Third World.[71] Rapid industrialization and exploitative trade relations were creating cycles of dependency, impoverishment, and environmental decline.

Conventional environmental protection has long been synergistic with modern development. For example, as national parks were set up throughout the American West, their landscapes became reformulated as sites for managed protection and in many cases for state-sanctioned resource extraction.[72] The national forest model was replicated throughout the world, and conservationists and preservationists alike viewed people as obstacles to environmental protection. It was widely believed that the goals of conservation were in direct opposition to the interests of local communities living in forests, leading to policies of resource management that required the strong arm of the state. This is perhaps best illustrated in Mark Dowie's book, *Conservation Refugees*, which details the displacement of formerly colonized Indigenous peoples from about half of the 108,000 officially protected conservation areas worldwide.[73]

Sustainable development has contributed to a form of greenwashing in which market-based solutions are privileged over other alternatives, benefiting some of the same players who contributed to environmental disruption in the first place. This has further excluded and marginalized disempowered communities who were disproportionately harmed by development policies. For the world's diverse and stratified population, the consequences of colonization and development have been uneven, leading to an increasingly uneven playing field for sustainable development. For many, modernization has translated into lived experiences that are marked by perpetual poverty and social upheaval—a reality that has simultaneously served as the quandary and the consequence of development thought and practice.[74]

Meanwhile, the culprits of environmental injustice are rarely held accountable. The project of globalization has created escape routes for the leaders of polluting industries, creating a convenient gulf between production, consumption, and disposal.[75] As the Indian scholar and feminist activist Vandana Shiva explains, globalization opened a political space in which the dominant could flee, "freeing itself from local, regional, and global sources of accountability arising from the imperatives of ecological sustainability and social justice."[76]

COMPETING LOGICS OF COASTAL PROTECTION

On the shore, competing logics of coastal protection, from "hard" engineering interventions like building more seawalls to "soft" measures like building or regenerating natural barriers, mirror the same inherent values and tensions associated with the political and institutional project of sustainable development. These approaches are also referred to more broadly as the transition from "gray" infrastructure to "green" infrastructure, reflecting the turn to environmental values following the ecological crisis of modern development.[77] In this context, placekeeping is now being negotiated across shifting values of coastal protection and adaptation. In Guyana and the Maldives, and throughout the world, seawalls have become boundaries around which these competing desires are being negotiated.

If gray infrastructure emerged in a context of resource extraction, the expansion of global capitalism, colonialism, and industrialization, green alternatives are the response to the failure and high cost of these structures. The gulf between hard and soft logics of coastal protection is driven by opposing political values, yet both logics emerge from a common goal of modifying or "improving" the environment for human uses. While these divergent paths entail different commitments and environmental values, they share the pursuit of protection and are both capable of dark possibilities, including forced relocation and outcomes that exacerbate mobility injustice, racial injustice, coastal injustice, and climate injustice.

Hard coastal protection measures include the construction of seawalls and other gray infrastructures to protect the coastal environment from hazards such as hurricanes, tsunamis, and sea change for economic, political, or cultural purposes. These measures are often accompanied by projects of land reclamation, which involve the large-scale draining of intertidal zones by pumping and sediment transfer. Land reclamation is in fact one of the earliest forms of coastal protection and has had an enormous impact on the expansion and development of coastal areas.[78] Reclaiming land from the sea often requires hard structures to anchor newly reclaimed land in place. As such, seawalls are a companion to land reclamation.

Closely connected to land reclamation is the construction of artificial islands, which similarly accompany the construction of seawalls. In their article "'Dreams So Big Only the Sea Can Hold Them,'" Mark Jackson and Veronica della Dora conceptualize artificial islands as sites of emergent social space emanating from the Western geographic imagination.[79] They differentiate between late modern "post-Dubai" fantasy islands and the more traditional artificial islands that have appeared across the world over the centuries in the form of oil platforms and other industrial landforms. The difference, they argue, is that these new islands are the "ultimate dream-spaces of global capitalism," where utopian worldmaking is envisioned and enacted by the wealthy and powerful.[80] As both material constructions and assemblages of thoughts and ideas not yet in physical form, artificial islands have come to embody the highest technological hopes for a future fully detached from the limits and constraints of the natural world.[81]

The turn to soft engineering came in the 1980s as industrialized nations shifted to a discourse of sustainable development. Coastal engineers similarly shifted their attention to environmental concerns. By the 1990s, soft measures for managing coastal erosion were gaining traction and popularity, promising to preserve the aesthetic of the natural coastline while also protecting the financial interests of property owners.[82] These projects claimed to work *with* nature to restore natural defenses by building up dunes and nourishing beaches in popular tourist locations. During the 1990s, beach nourishment became one of the most common trends in coastal defense. Beach nourishment aims to both protect coastal infrastructure and maintain the aesthetic of a natural beach—but not without its social and environmental consequences.

Soft pathways include the construction of green infrastructure and projects of regeneration, including efforts to restore and re-create past ecosystems. This entails actions such as planting mangrove seedlings to bring back a protective green belt along a coast, referred to as "the green seawall" in Guyana, or planting coral barriers to protect fragile island ecosystems, as is the case in the Maldives. These sorts of practices, as Emma Marris sees them in her book *Rambunctious Garden: Saving Nature in a Post-Wild World*, produce "designer ecosystems" that are based on a vision of premodern contact, driven by a reimagining of the existing baselines of human disruption and a desire to reconnect with pre-anthropogenic forms of protection.[83]

In the context of rising sea levels, advocates for soft measures represent a new and more conservationist school of thought, emphasizing green infrastructure while questioning the economic, ecological, and aesthetic logics of armoring, insuring, and rebuilding vulnerable coastal areas. Green infrastructure projects are increasingly placed under the banner of "nature-based solutions," a term that is used across many research communities in sustainable development, urban green-space planning, and ecosystem services, as well as hazards and vulnerability reduction.[84] The term emphasizes actions that mimic nature to cope with environmental extremes. As such, it has much in common with logics of ecological resilience, premised on what critics refer to as "coerced resilience," where resilience is supplied through artificial maintenance rather than by the ecological system itself.[85] David Chandler argues, "Even if this could be achieved, 'natural processes' would be further modified by anthropogenic manipulation: the mere need to intervene to 'bolster' these allegedly 'natural processes' would inevitably produce other unintended stresses and strains."[86]

In contrast, advocates for seawalls and other hard measures represent the old guard; they prioritize gray infrastructure and question the social and cultural losses that are implied by the choice to move infrastructure in order to regenerate vulnerable areas, arguing instead for the importance of protecting major cities and sites of cultural and economic significance. Like other large-scale development projects, seawalls have been promoted as a means to eliminate poverty and to create a better life for the world's poor who occupy difficult environments. Not surprisingly, the justification for hard engineering projects is now framed in the language of adaptation, under the dubious claim that they act as barriers to climate change.[87]

While much of the discourse on hard and soft coastal protection measures is driven by a concern over the projected impacts of climate change, other drivers and motivations are often hidden. The extent to which coastal protection is actually intended to address climate change or to serve the purposes of development and the financial interests of the wealthy is a question that requires deeper analysis. In the contemporary context of climate change, land reclamation and the construction of artificial islands are increasingly framed as adaptation measures and sometimes as alternatives to managed retreat.[88] In the Maldives, for example, converting lagoons into solid ground enables new housing schemes and economic opportunities, but it also introduces many new problems and environmental risks. Not only does land reclamation reduce resilience by disrupting natural cycles; it also creates path dependency that locks future governments into maintaining these projects indefinitely, often by means of seawalls.[89]

Nature itself is a site of power, and in many ways, regeneration involves negotiating with the disruption of the past alongside the same actors that drove the modern project into foreign lands in the first place. The baselines of regeneration are largely articulated by those who have the privilege to define them. As William Cronon cautions, the notion of wilderness as pristine and untouched is a fantasy.[90] This is particularly troublesome in contexts where the definitions of "local heritage" are defined not by Indigenous inhabitants but by nonlocal intervenors, whose interventions, despite good intentions, echo the uncomfortable dynamics of settler colonialism.[91] First World environmentalisms are largely concerned with the preservation of a pristine vision of nature, characterized by the work of Ralph Waldo Emerson and Henry David Thoreau.[92] In contrast, the environmentalisms of the poor are rooted in clashes over productive resources interlinked with issues of distributive justice (the fair allocation of resources) and human rights.[93]

Regeneration is a product of the modern project. According to Wolfgang Sachs, "Regeneration takes into account that the royal road of development has vanished."[94] This observation is also made by Anna Tsing and David Chandler, who consider how regeneration is perhaps the final attempt of the powerful to apply technocratic approaches "to short-cut problems rather than to tackle them at the source."[95] In this way, regeneration represents the "ecomodernist fantasy" of the "good Anthropocene."[96] Nature-based solutions overlook the roots of ecological disruption, opting instead for managerial strategies and improved resource efficiency.

How such transitions affect vulnerable communities is a question that remains tied to existing concerns about urban greening, green climate gentrification, and the displacement of racialized and minoritized people, as well as social and cultural exclusion.[97]

As adaptation planners shift their gaze from gray infrastructure to green infrastructure, the underlying logics that drive these interventions are left unquestioned. There are many instances where the boundaries are quite blurry. While green measures are often presented as altruistic and enlightened understandings of natural systems, it is important to recognize that the shift came in response to the wholesale destruction of the coast through the development of hard structures. Coastal modification has contributed to the widespread problem of coastal erosion as well as artificial systems of value placed on shoreline aesthetics and assets.

When faced with the very real prospect of catastrophic sea change, most planners recognize that a combination of gray and green infrastructure is necessary, and as such, the emerging geographies of sea level rise will be shaped by both old and new forms of coastal stabilization. However, the manner in which a pathway is selected or rejected is deeply political, and the resulting physical structures will inevitably embody the politics of their design and maintenance. Throughout the process of imagining, creating, and maintaining coastal protection systems, the people who live with coastal vulnerability are rarely asked which option they prefer. Rather, coastal adaptation is usually managed from the top down, by government agencies with economic agendas. Furthermore, each pathway enacts social and environmental ripple effects with potential injustices for neighboring communities and future generations.

CONCLUSION

Never fully washed out to sea, the past remains embedded in coastal landscapes, narrowing, and in some cases swallowing, future escape routes. Placekeeping is predicated on the modern project and the vast and incalculable environmental damage of coastal disruption. Today, efforts to stay in place are caught between competing desires for gray infrastructure and desires for nature-based solutions and green infrastructure. On the shore, placekeeping engages with complex entanglements of infrastructural systems and coastal engineering practices that emerged as companions to the drivers of global inequality as well as in response to the damages these transformations initiated. As such, coastal protection occurs in the context of human politics, in particular, relationships of power, conquest, expansion, and, more recently, regeneration. From this vantage point, seawalls and other coastal protection measures can be understood as socially constructed technologies entangled in the making of empire. They are also bound in values that are shaping the modern world through material manifestations that play a profound role in the making and remaking of place.

The complex and ever-shifting cultural meanings of seawalls come with specific social and environmental implications, which are examined further in the cases of the Maldives and Guyana. While seawalls are eventually grounded in place, they are in fact always in motion and undergoing ideological struggles. The coastal infrastructure of maritime trade and agricultural irrigation traveled to the colonies. Dutch seawalls found their way to Guyana. French seawalls found their way to Japan, and Japanese seawalls found their way to the Maldives. The next two chapters look specifically to Guyana, where seawalls work in tandem with soft engineering designs to distribute protection to people and areas of the shore that are deemed vulnerable. Through the story of Guyana, we are able to see how seawalls are ideologically and physically embedded in the politics of placekeeping, entangled in attachments to place that are being negotiated alongside histories of injustice.

2 The Strangled Shore

Guyana's shore is the subject of an obscure film titled *The Seawall: Tales of the Guyana Coast.*[1] It follows a character referred to as the Old Man of the Wall, played by the Guyanese actor Mark Matthews, who is granted a magical and timeless presence as he floats between documentary storytelling, historical reenactments, and fictional vignettes. He takes up residence near the seawall as a bystander and later as a squatter. His narrative, written by Rupert Roopnaraine, flows like a stream of consciousness, carving through space and time like Guyana's many rivers.

The story begins with a suicide. A young man runs along the foreshore and climbs up the side of the seawall to gather stones from its cracked surface. As he looks into the sea with hypnotic disconnection, he fills his shirt with the stones and submerges himself in the Atlantic. The scene then cuts to images of Guyana's shore and waves slashing against seawall after seawall. The Old Man of the Wall somberly states, "This water washes over magic, murder, and enslavement. Between we and this ocean we have a wall drawing a line between an ocean of land and an ocean of water. And when the wash come, it can break."

Images of seawall repair follow. The old man hovers over aerial footage of the coast to the sound of a plane's propellers. With a weary expression, he mutters, "Once more, an ancient battle joined, again and again, from time immemorial. Today's crack is tomorrow's breach. All along the coast the wall's crumbling from neglect." The ocean's transcendent force is felt widely throughout the coast, and the Old Man of the Wall cautions, "When the beast has gone back to sleep, that is the time when man can come out again to harvest what the sea has left behind. To worship and carry on the daily struggle of survival."

The film then shifts back in time to a plantation house, and the color fades to black-and-white. In this alternative reality, the Old Man of the Wall is enslaved to a colonial family that speaks at length of efforts to construct sea defense structures. The fictional reenactment goes to great lengths to show how difficult and expensive sea defense was for colonizers. The film proceeds to move back and forth in time, revealing the ruins of empire and the abandonment of colonial structures for liberated and impoverished Guyanese to contend with. One is left with the feeling that the seawall is the physical embodiment of historical oppression—a companion to a people who are themselves drowning, submerged by their own rocks.

At the end of the film, the old man muses once again. His voice rises over a crackling campfire, calypso sounds, and dim footage of the Atlantic Ocean. "If only this wall could talk," he starts. His words echo over nostalgic images and happier times as he speaks of the good old days when the wall felt "free and easy" and was "busy up and down, driving, gathering, walking, politics in the air." Then his mood shifts as he recalls the reality of the present moment. "Sometimes it's like the sea is calling for sacrifices, as if the sea made this wall into an altar," he says. "What brings us here to this wall of dreams, of nightmares, mysteries, murders, and strange, strange happenings? What ancient memories bind us here where we ceremony and we worship?"

In Guyana, coastal protection is grounded in colonial dispossession and extractive capitalism. Throughout Guyana's modern history, seawalls have been tried and tested across a variety of political landscapes, from colonial rule through independence and US interventionism, culminating in a complicated sense of place and a dependency on concrete. As oppression and place attachment are co-constructed through projects of dispossession, efforts to stay in place become embattled in the social and material assemblages of seawalls. The construction of seawalls has materialized an inherent politics of colonial experimentation and authoritarian governance. Now, as sections of the seawall fall into disrepair, these politics are being renegotiated and challenged.

What follows is a closer look at the making of Guyana's extensive network of seawalls and coastal defense structures, bringing together a series of events and relationships centered on processes of colonization, reclamation, maintenance, and a nostalgia for the past. In order to grasp current efforts to restore seawalls in Guyana in the face of climate change, it is necessary to trace their origins. Over hundreds of years, practices of coastal stabilization have worked to extract resources through digging, pumping, and dredging, disciplining the physical landscape of Guyana. These practices have also had an impact on the social and cultural landscape of Guyanese identity. It is impossible to understand coastal Guyana without making reference to this past; the bitter traces are everywhere, inscribed in the very landscape. Scratching through layers of residue left behind by Dutch, British, and Guyanese efforts to control the hydrological features of the landscape, new understandings of current struggles to adapt to sea change become visible.

THE ORIGINS OF GUYANA'S SEAWALL

When I first arrived in Guyana to understand the existential dilemma of seawalls, I met with a geologist who at that time was Guyana's acting minister of public infrastructure. As I entered his air-conditioned office with my recording device in hand, I was immediately struck by the antique maps covering the walls that depicted various stages of Guyana's cartographic history, a stark contrast to the technical blueprints that I regularly saw scattered about in other offices.

Prompted by the maps, I asked about his earliest memories of the coast. He closed his eyes for a long moment before responding, and when he spoke, his words came slowly. "There were no sea defenses when the Dutch came," he began. "We started to settle on the coast and polder during European colonization. At that time, most of the Amerindians had died off or had been killed off, either deliberately or by disease." The story he soon unfolded, like the cartographic story of the old maps on the wall, was one of colonial disruption and a shoreline where the past refuses to surrender and the future refuses to wait. It became clear that the map of the future is unfolding in social and ecological contexts that are historically entrenched and inherently unjust.

Colonial land reclamation, along with its early seawalls, squeezed and trapped the shoreline, setting in motion a series of events that would drive the country into a state of economic ruin and racial divide. Upon reflecting on the difficulties of maintaining the seawall, one coastal engineer told me that the Dutch had literally "strangled the shore." He repeated the sentiment a second time for emphasis: "They *strangled the shore* by building these sea dams, canals, and sluices."

In the beginning of the seventeenth century, Dutch colonizers settled inland on the Demerara River, calling it "the Wild Coast."[2] At that moment in geologic time, the coastal plain appeared stable, with the ground slightly elevated and subject to flooding only occasionally during the lunar high spring tides.[3] Yet, as the Dutch would later discover, the foreshore was in fact moving.[4] Over the course of the Holocene, massive deposits of mud from the Amazon River had accumulated along the coastal plain, patches of which were in slow motion cycles of erosion and accretion. Multitudes of black, red, and white mangroves had rooted into the foreshore, holding the mud in place while dampening the titanic force of the Atlantic Ocean.[5] During this time, the Dutch were well acquainted with sea defense and coastal reclamation, having brought with them a set of skills harvested during a long and bitter history of erosion, accretion, reclamation, and submergence.

Johan van Veen reflects on how the Dutch came to capture land from the sea in *Drain Dredge Reclaim: The Art of a Nation.*[6] He describes the primordial terrain of the Netherlands as a repository of sand and mud washed out during the closing act of the Ice Age. Initially, few ventured into the marshy waters that gave birth to the low-lying Dutch coast. The tribes that did, later known as Frisians, watermen, mud workers, and dike masters, kept themselves above water by building artificial mounds, removing one hundred million tons of clay earth by hand.[7] Pliny the Elder encountered these marsh villages as he took part in the Roman conquest of the Chauci along the Rhine River. He documented his findings in *Naturalis Historia* under the poetic title, "Countries That Have No Trees," noting that the landscape "opens a question that is eternally proposed to us by Nature, whether these regions are to be looked upon as belonging to the land, or whether as forming a portion of the sea?"[8]

The Frisians were among the first to notice that exposed marine sediment provided a particularly productive base for agriculture. In fact, this discovery was so important that a politics of reclamation and ownership emerged, around which battles and civil wars were fought.[9] In medieval Frisia, the necessity of maintenance justified the decreeing of laws that became as harsh and as unforgiving as the sea itself. To ensure that every man did his share of the work, "dike or depart" became the rule of the land. Anyone who was unable to help at the dikes had to leave the country, and those who refused to help were put to death, sometimes buried alive in the very spot that was breached by the sea. Furthermore, if a man tried but failed to seal a breach, the "law of the spade" would allow a rival farmer to take his land.[10]

For the Frisians and their ancestors, the construction and maintenance of seawalls was a matter of utmost importance, next to freedom. They fortified their coastline with a "Golden Hoop" of seawalls. These early structures survived hundreds of years of political conflict and struggle, creating a foundation for the contemporary nation of the Netherlands. Over time, the construction of dikes to protect against flooding enabled more intensive agricultural practices and a higher demand for coastal protection against the increasing likelihood of floods.

Embedded in the ever-changing Frisian landscape were hardships that gave credence to the Latin saying, "Frisia non cantat" (the low part of the Netherlands does not sing).[11] As seawalls multiplied and polders grew larger, Dutch populations grew with them, and breaches became more catastrophic. The relationship between drainage and disaster was paradoxical. As Van Veen explains, "The more we raised the dikes and the better we drained the country, the higher and more destructive became the floods."[12] Tragedy often struck without warning, as was the case on the evening of November 18, 1421, when sixty-five villages and ten thousand souls were consumed by the sea. The catastrophe claimed over 3 percent of the population and was due in large part to the creation of a thirty-mile polder—a rectangular piece of land reclaimed from the sea and rivers with dikes and seawalls—too bold an experiment, perhaps, for such an early date.[13]

The dike masters built on this tragedy and many others that followed. In one drowned landscape after another, they developed their ability to "strangle wild streams" while handling "the unwieldy dredging machines with unerring skill and in the cheapest possible way."[14] And while tragedy often started with a breach, the sea was at times creative in its methods of destruction. For example, in 1730, an outbreak of wood-boring sea worms invaded the defenses along the Dutch coast. An apocalypse was in the making, so much so that "churches ordained public prayers for the removal of this 'terrible scourge of God.'"[15] *Teredo navalis,* a naval shipworm found in temperate and tropical seas, had tunneled into the hulls of vessels returning to the Rhine Delta. The terror was captured in a 1731 print by Abraham Zeeman that exaggerated the size of the creatures, accentuating "the monstrous interpretation of this invasive species."[16] Dike masters worked frantically to remove and replace elaborate systems of wooden pilings with stone.[17]

As the Dutch exerted dominance over the whims of the intertidal zone at home, they expanded their dominance to other parts of the world, including Guyana. In 1738, Laurens Storm van's Gravesande was sent to Guyana by the Dutch West India Company to colonize the mud-clad coastland. In contrast to their upstream trading posts, the soils of the lowlands were rich in nutrients and ripe for sugarcane. All that was needed, as it occurred to these future planters, was to tame the shore by strangling the high spring tide. Storm van's Gravesande's first project was to rebuild the wooden military fort at the mouth of the Essequibo River with brick and mortar, removing any possibility that a *T. navalis* invasion would undermine Dutch control of the waterfront.

In Guyana, the construction of Fort Zeelandia, as it came to be known, was to safeguard the interests of the Company from the social instabilities of Europe and the threat of neighboring French and British colonies. On completing the project, Storm van's Gravesande then opened the Demerara coast to English planters from the British West Indies, offering tax breaks and free parcels of land. The map of Guyana's coastland was indexed and carved into pieces that protruded inland perpendicular to the coast.[18] The shape was dictated by irrigation canals, a necessity for the cultivation of sugarcane.

The construction of sugar plantations called for an almost complete reconfiguration of the landscape, starting with an investment in sea defense and reclamation. Captain Edward Johnson surveyed Guyana and the Demerara coast at the time of early development and described the scene as "the most remarkable of any part of the Coast; the Woods, in many places, being burnt down and cleared for cultivation."[19] Each new estate required a system of seawalls to prevent flooding at high spring tide, back dams to block rainwater from the lands behind the estate, side dams to block floodwaters from the neighboring lands, and multiple drainage canals with sluices (also called kokers) to release surplus water from the estate.[20] All of this had to be constructed before a single crop could be harvested.

The building of dams, seawalls, sluices, and twelve-foot-wide canals was made possible by enslaved labor. In *The Making of Guyana*, Vere T. Daly envisions how Abomie, the slave of Storm van's Gravesande, might have described the difficulties of sea defense in Guyana: "He would without doubt tell of the hard work which the slaves had to do."[21] According to one estimate, forty-nine miles of canals and ditches were needed, in addition to sixteen miles of drainage canals for each square mile of sugarcane cultivation.[22] The original plantations on the coastal plain necessitated the removal of approximately one hundred million tons of earth.[23] Remarkably, this is the same amount of earth estimated to have been removed by the Frisians for the formation of mounds in the early Dutch region.[24] However, unlike the Frisian dike masters, the labor of enslaved people in Guyana is hardly recognized in history books.

As conflict rippled throughout Europe, British planters competed for control of the sugar economy. The Netherlands lost its monopoly in 1814, and by 1831, the sugar estates of Guyana were unified into British Guiana. However, Dutch methods of sea defense were not soon forgotten. As the cultivation of sugarcane intensified under British rule, "Dutch experience was transferred to the Guianas . . . to carve out individual but contiguous plantations on the coastal plain."[25] The Slave Trade Act of 1807 and the subsequent emancipation of enslaved people in 1838 caused a shortage of labor, slowing colonial development in British Guiana. Indentured servants from India were transplanted into the Guyanese workforce to dredge the existing canals and work the cane fields. The thick claylike mud posed perpetual problems for drainage, and without access to the expensive machinery found in the wealthier nations, dredging continued to be done by hand.

GUYANA'S PLANTATION COAST AS A LABORATORY OF COLONIAL ENGINEERING

The making of Guyana's coast is embedded in waves of injustice and systems of plantation capitalism that include colonial experimentation. In *A History of the Guyanese Working People, 1881–1905*, Walter Rodney analyzes the origins of plantation capitalism in Guyana and the material constraints of sea defense on the country's development. The "planter class," as Rodney calls it, had the largest economic stake in sea defenses (142,000 acres in 1880 divided among 113 sugar estates).[26] Colonial development remained almost exclusively tailored to the exploitation of the coastland, and despite the presence of minerals inland, including timber, gold, and diamonds, there was not a concerted effort to shift economic activity away from the coast. This was due in large part to the interests of the planter class and the politics of sea defense. As Rodney argues, "Planters viewed the hinterland as a potential competitor for labor and sought to dissuade freed blacks from moving in that direction . . . just as they had earlier sought to forestall slaves from escaping into the bush."[27]

Despite the many political advantages commandeered by the planter class, their piecemeal efforts to hold back the sea were largely futile. The direction of sea defense had led to the uneven development of the coast while contributing to the wider problem of polderization, whereby the coast was collectively sinking due to the expanding layout of contiguous plantations. For wealthy planters, erosion was a compulsory burden that intersected with a host of economic challenges. The main concern for planters was the prospect of drowning in the costs of repair. In the *Argosy*, a newsletter shared among British planters, proprietors, and attorneys, the plight of sugar cultivation was widely recognized as a paradox: "Report after report on the sugar industry in British Guiana returned with some astonishment to the unique problems of the sea in front and the savannah floods behind and of the need to simultaneously exclude and conserve water."[28]

Historical maps of the Demerara coastland outline rows upon rows of phantom plantations.[29] As Rodney notes, planters often shifted their priorities, sacrificing segments of their plantations. "If it proved economically or technically impossible to continue repairing a front dam in a given location," he writes, "then the estate 'retired' that dam many roods inland and renewed the struggle after conceding valuable acres of its frontlands."[30] Rodney explains that "when an estate was said to have 'gone under,' this was more than a merely figurative expression, because it was usually the invading sea that completed the demise of a failing estate."[31]

To put an end to the intermittent cycle of erosion and retreat, private colonial planters schemed for public funds to connect and bolster the seawalls along the coast. A thirty-five-year debate ensued as divergent interests stymied any resolution on public oversight of sea defense. Then in February 1855, disaster struck. Governor Philip Edmond Wodehouse, in correspondence with the British House of Commons wrote, "On the afternoon of the 17th, the sea rose during spring tide to a height, and with a violence unknown for nearly fifty years, and, in the course of a few hours, swept away nearly the whole of the upper part of the embankment."[32] The high tide continued to inundate the capital city of Georgetown: "The public road was flooded twice a day; the sea thundered over and through the breaches; the city was in danger."[33]

In an effort to drive back the water, the government strengthened the seawall at Georgetown, extending it for two miles with convict labor from the newly established penal settlement on the Mazaruni River. Two years before the flood, the convicts had been shipped from the neighboring British West Indies to work in the granite quarry near the Mazaruni penal settlement. James Rodway, a contributor to the *Argosy*, wrote of the haphazard way in which the sea defenses were erected: "Thousands and thousands of tons of stone were thrown along a line."[34] In this fashion, the shoreline was armored.

In 1883, the legislature formally assumed responsibility for governing sea defense. The interests of the major sugar plantations were often expressed as in the interest of the public, their owners claiming that the protection of large estates was in line with the well-being of the masses. Meanwhile, the needs and modest demands of villagers and small farmers were systematically ignored. Rodney recounts how colonial planters engaged in victim blaming to justify their continued domination of the coastal landscape:

> It was common practice for planters and their officials to scoff at the villagers and to interpret their troubles as having derived from their own inherent

> inadequacies. . . . Only the cultivation of sugarcane was considered industrious by the planters, while peasant production of provisions was said to have been a waste of time and resources. This attitude strongly influenced planter decisions on most of the questions connected with sea defense, drainage, and irrigation.[35]

As the funding for sea defense structures increasingly shifted from the responsibility of private planters to the government, plantations took advantage of public loans to manage their drainage and sea defense problems, inviting conflict and disparity along the coast. In 1906, the Sea Defense Commission was formed, chaired by the director of public works, to oversee sea defense projects on the east and west Demerara coasts, and by 1913, the government was covering one-fourth of the cost of coastal fortification.

In search of a permanent sea defense scheme, the planter class turned its attention to debating the most effective measures. At the time, there was little consensus: "Some planters favored fascine dikes, others relied on the increasing use of piled granite boulders, and still others were partisans of groyne construction."[36] Groins are structures positioned perpendicular to the shore to gather sediment and manage erosion.[37] Vertical sheet piling was also a common fixture along sections of the coast. Attempting to settle disputes among planters, the Colonial Office commissioned the Dutch engineer and coastal expert Baron Siccama to determine the best method of sea defense in Guyana. In a report submitted to the Department of Public Works in 1891, Siccama advised the colony to continue to build stone seawalls in areas prone to erosion.[38]

Meanwhile, an alternative method of sea defense was being tested in the village of Dymchurch, located in the wetlands of southeast England along the reclaimed region of Romney Marsh.[39] Edward Case, who had spent his career as a civil engineer for the Public Works Department of Ceylon (present-day Sri Lanka), was appointed in 1890 to oversee sea defenses in the Romney Marsh area. By 1892, the Dymchurch seawall was in dangerous condition after a difficult winter had destroyed large tracts of the stone-pitched slope, leaving the wetlands in a state of jeopardy. In a survey of the damage, Case determined that the seawall was the leading cause of erosion, an observation that went against the logic of the time.[40] The wall, originally an earthen embankment lined with stone, was paved in 1822, altering the natural foreshore. Each time the wall failed, the authorities would repave the damaged sections, draining resources. Rather than build stronger pavement along the seawall, Case wanted to work with the shape of the foreshore. He wrote, "It is the business of the engineer to utilise the forces of nature, and not to oppose them."[41]

The Case System, as it came to be known, was marketed in London and applied in Dymchurch and other areas. Cecil Case and Frank Gray helped implement the vision of Edward Case, building sea defenses "utilizing the forces of nature to build up a natural protection instead of fighting the sea."[42] The method involved using low-lying degree angles to the shoreline, a system designed to operate as a machine of accretion, trapping sand and allowing the shore to build up along the outer facing wall of an existing embankment. When Edward Case suffered an untimely death in 1899, his obituary read, "This system of groyning, with which Mr. Case's name will be chiefly associated, is remarkable for its simplicity of design, the economy it effects in time, labour and cost of materials, and its successful results at Dymchurch and elsewhere."[43]

In British Guiana, the Case System was under experimentation, but the results were largely unsuccessful. According to one report, groins installed in 1899 were insufficient in trapping sediment, causing erosion instead.[44] Gerald Otley Case, the son of Edward Case, became a consulting engineer for British Guiana in 1916. On seeing the existing patchwork of sea defenses in the colony, Gerald Case reported that "the rapid erosion in recent years is to a large extent due to artificial causes arising from the empoldering of the marsh land and the construction of badly designed sea defense works."[45] He stressed the need to determine more appropriate angles and heights for the next generation of sea defense structures and the importance of maintaining newly installed infrastructure.[46] Resolved to make the Case System work in Guyana, Gerald Case conducted further experiments, adjusting the angle of the groins, paving the way for modern sea defense in Guyana.

As Gerald Case convinced the director of public works in British Guiana to work *with* the forces of the sea, the coastline began to look increasingly modern, with concrete fixtures. Starting in 1916, an extensive network of reinforced concrete groins was installed along fifty-two miles of the Essequibo and Demerara coasts. In addition to the 111 groins installed along the shore, an additional ten miles of seawalls and sluices were reinforced with concrete and "Lignocrete"—a combination of sawdust and cement—Gerald Case's invention "to cheapen the cost of construction."[47] To expedite the process, a special arrangement was made between the United States and the British government to grant priority permits for shipping construction materials.[48] The initial work involved the demolition of many of the old stone sea defenses, and by the end of 1919, "approximately 1,500,000 cubic feet of stone alone had been removed from wave screens on the foreshore of the East, West, and Essequibo coasts."[49] New sections of the seawall were installed based on the Case System principle of building with "such a slope that waves can harmlessly roll up the shore and expend their energy before reaching the high water-line."[50]

As the work neared completion, Gerald Case was pleased with the outcome, reporting to the governor of Guyana, "After 30 years' experience in sea defence works I am of the opinion that nowhere in the world can be found a better example of the efficiency of groynes than the East Coast of Demerara."[51] However, the results of the Case System were uneven as some of the groins installed along the west woast of Demerara were quickly starting to fail. To maintain these newly developed structures, the government launched the Sea Defense Ordinance in 1933, granting a twelve-member Sea Defense Board the power to levy taxes to pay for breaches without waiting for legislative approval.

On January 7, 1934, a combination of heavy rainfall and strong winds launched the coast into a state of emergency. According to one historical account of the breach:

> The incessant rainfall combined with breaches in the sea defences and the Lamaha Conservancy dam further compounded the situation leading to extensive flooding along the entire coastal strip. . . . It was reported that residents along the East Bank were forced to seek refuge among the rafters of their homes as they were trapped, surrounded by floodwaters over three feet high. . . . The newspapers reported that thousands of coconuts were floating down the river and cattle, pigs and fowls were drowning.[52]

During the floods, a large portion of the seawall protecting Nog Eens washed ashore ("Nog Eens" is Dutch for "Try Again"). Overall, the damage was so costly that the governor commissioned a committee to investigate its causes. The Flood Investigation Committee "noted that the lack of maintenance by derelict or partially derelict estates contributed significantly to the flooding of villages as dams were not cleared in an 'efficient manner.'"[53] The sugar industry, having lost an estimated 20,127 tons of yield, was given a subsidy for the reestablishment of its fields.

The design of the Case System was primarily focused on the sea side of British Guiana, ignoring the larger system of irrigation for sugarcane production, particularly the reservoirs, canals, and back dams. Furthermore, Case failed to account for the dynamics of Guyana's muddy foreshore, which was not stable but in motion, eroding and secreting in thirty-year cycles. More than a century later, one British coastal engineer observed that these early engineers "who built up expertise based on intelligent observation" were sometimes successful but limited by a lack of vision "to take in the complexity of tidal and river flows, sediment transport and wave action and to predict how estuaries would respond to works."[54] The necessary data and tools would not emerge until the end of the nineteenth century.

INDEPENDENCE, A SHORT-LIVED REVOLUTION, AND BURNHAM'S WALLS

Polder by polder, the patchy architecture of sea dams enclosed the coastal plain of Guyana, marking the physical boundaries of production as well as the social boundaries of upward mobility. After the abolishment of slavery, as people moved into towns and villages, some purchased abandoned estates by pooling their resources. However, small-scale farmers were unable to compete with the political and economic resources of wealthy planters. According to Walter Rodney, "All the workers on the coast and in the interior remained under the shadow of the plantation."[55]

Free African and Indian wage earners, Indian indentured servants, and peasant farmers could not manage the maintenance costs associated with the upkeep of seawalls, dams, and canals and were therefore locked into poverty. As Rodney explains, "Villagers were not simply the victims of 'natural' environmental forces[;] . . . rather as workers and peasants, they were doubly disadvantaged by the social and political inequality of the colonial capitalist system."[56] The common people were prevented from achieving economic success. Cultivating alternative crops could not yield the needed profits for canal maintenance and sea defense. As Rodney points out, "Wealthy landowners barely managed to improve their own drainage, irrigation, and sea defenses; while poorer proprietors often sold out or abandoned cultivation because of expenses under these heads."[57]

By the 1940s, as economic depression hit British Guiana, public perception of the government became more cynical. In his 1939 book, *Labour in the West Indies: The Birth of a Workers' Movement*, Arthur Lewis writes, "The impression is now widespread among the people that the Governor and Officials are little more than the tools of a white oligarchy of planters, merchants, and bankers, in whose society they spend most of their time and whose will it is that the policy of the government is the policy of the local club decided on perhaps, over a round of golf or a whisky and soda."[58] As the planter class struggled with the repeated collapse of its sea defenses, a Marxist revolution was in the making. World War II had crippled the British Empire, creating an opportunity for representatives of the Guianese working class to engage in what they referred to as a "politics of protest." In 1946, two left-leaning groups were formed. Cheddi Jagan, the offspring of indentured laborers from India, cofounded the Political Affairs Committee and contested a seat on the legislature, winning by 31 percent of the vote. Forbes Burnham, Jung Bahadur Singh, J. A. Nicholson, Hubert Nathaniel Critchlow, and Ashton Chase formed the British Guiana Labour Party and won five seats. By 1950, the two groups merged, creating the People's Progressive Party (PPP).

At the head of the PPP was Cheddi Jagan, a vocal advocate for universal suffrage, self-determination, education, land reform, and "a complete system of diversified agriculture."[59] Jagan sought to bring an end to British rule in Guyana and build a socialist society in its place. In 1952, when the British Colonial Office sent a commission to examine proposals for constitutional reform, Jagan protested the commission's plan to implement a slow and incremental pathway to independence. In a passionate response, Jagan argued that it was a scheme to perpetuate the rule of the planter class:

> Maybe we do not have bloodshed and the roaring thunder of the guns and the whip-lash of the slave-owners, but, nevertheless, we have the influence of sugar in this country so protected—so strategically placed—that the farmer, the wage-earner, whether in this industry or out of it, is nevertheless in its clutches.... It suits the policy of the sugar industry, however, not to introduce land reform in this Colony because the moment any vital land reform is introduced here ... the farmers of the country would be afforded an opportunity to have a square deal.... There is no doubt about it that the present system is outworn and no longer satisfies the wishes of the people, because people are fed up with the way in which the country has been governed.[60]

To the shock of the government, the PPP won eighteen of the twenty-four elected seats in the House of Assembly during the 1953 general elections, elevating Jagan to the position of chief minister. Jagan's term lasted a mere 133 days, ending with the arrival of British warships in the port of Georgetown. The British, nudged by the demands of the United States, claimed that the PPP was a "communist organization" allied with the Soviet Union and used martial law to subvert the party.[61] Jagan and other elected officials were deposed and imprisoned for civil disobedience. Five months behind bars awakened an ephemeral poet in Jagan, prompting him to write "Death to Imperialism," a poem that speaks to the hypocrisies of the new world order and ends with the line, "Our beautiful country, a vast prison you have made."[62]

Undeterred, Jagan went on to win control of the colonial legislature in 1957 and again in 1961, but the vicious circle of oppression would continue, hastened by the divisiveness of racialized colonial politics and the erosion of socialist possibilities in the ideological battle of the Cold War. The initial removal of the PPP, later termed an "imperial *coup d'état*," undermined the chance for an ethnically unified road to self-determination. Cheddi Jagan's wife, Janet, who was also involved in politics, wrote that the somber events of 1953 were an attempt to "once and for all, destroy the PPP and terrify and intimidate the awakened people."[63]

Local entrepreneurs and outside forces opposed to the socialist leanings of the People's Progressive Party threw their support behind Forbes Burnham, a longtime rival of Jagan. Burnham, formally a member of the PPP, grew disgruntled when he lost a bid for leadership of the party. In 1955, he withdrew his membership and created the People's National

Congress (PNC), taking a faction of his supporters with him. As he competed with Jagan for control of the legislature, the parties began to split along racial lines inscribed by colonial divisions of labor. Burnham pitted urban Afro-Guyanese against the predominantly Indo-Guyanese farmers and sugar workers, staging a slew of violent strikes.[64] This set in motion a series of interracial conflicts so pervasive that a fellow founder of the PNC, D. J. Taitt, publicly stepped down. In a letter to the press, Taitt warned that Burnham was leading the country "into a blind alley of improvised tribalism at variance with the social and economic realities of the two major ethnic groups of our country."[65]

The multipronged reach of empire continued through US interventionism. The United States padded Burnham's pockets, hoping to sway the soon-to-be independent nation away from socialism and guide it in the direction of capitalism. Richard Lehman, deputy director of current intelligence, wrote, "If we do not put up the money we will eventually be faced—but probably not for three or four years—with an English-speaking Communist state in this hemisphere."[66] With outside support, Burnham succeeded in sweeping the 1964 general elections, and when Guyana became an independent nation in 1966 (changing its name from British Guiana to Guyana), Burnham and the PNC dominated the political scene for the next twenty years.

Spearheaded by the World Bank, the 1960s and 1970s became a period of yet another extensive rebuilding of the sea defenses along the east and west coasts of Demerara. During this time, great emphasis was placed on studying the hydrological dynamics of the foreshore in order to predict which areas would be affected by cyclical erosion. The Delft Hydraulics Lab conducted a study of the Demerara coast from 1960 to 1962, paid for by the United Nations Development Programme and the World Bank. The Delft report observed that groins "have little effect on foreshore conditions in the project area where the material is mainly clay, as the erosion in the lee of the current may undermine the structure."[67] Furthermore, the study identified the presence of "sling-mud" and predicted the erosion cycle of any given area along the Demerara coast is approximately thirty years.[68]

In 1965, the US Army Corps of Engineers released a technical review of the sea defenses with suggested revisions. Three designs were proposed based on the varying conditions of the coast. For the heavy conditions of the east coast, a wave screen was suggested, fortified at the bottom with gabions (baskets of plastic-coated steel mesh wired together before filling with stones) imported from Italy at a cost of $1.7 million per mile.[69] For the less severe areas of the east coast, it was suggested that a wave screen without gabion baskets would provide protection at a cost of $1 million per mile. Finally, for the west coast, a structure was proposed consisting of neither wave screens nor gabions but additional bulkheads to secure the existing revetments at $0.8 million per mile. According to Guyana's National Development Strategy, "These designs, different from Siccama's concepts, were ambitious and costly, but provided the sophistication required for an emerging nationhood."[70] Burnham's government borrowed an initial $5 million for the first phase of a twelve-year construction plan and another $5 million for the rehabilitation of Georgetown's seawalls.[71]

As Burnham built up the sea defenses of Guyana, he bolstered his authority by increasing the military budget and cracking down on his political opponents. The building contracts for the sea defense works were distributed according to the proclivities of the PNC government, and Burnham rewarded his supporters. He also allocated housing lots next to the newly rebuilt seawalls, coveted as beachfront property. While he oversaw the construction of Guyana's latest sea defenses, Burhnam approached his role in colonial style. In *Guyana: A Nation in Transit*, Ashton Chase, a founding member of the People's Progressive Party, describes how Burnham "would strut around on his horse, stirrup, whip et al, from one field to another," overseeing workers clearing trenches, drains, and canals.[72]

In 1974, Walter Rodney, the well-known and respected Marxist historian, spearheaded the Working People's Alliance (WPA) against the hypocrisy and tyranny of the PNC. The WPA became a political party in 1979, and in a speech made at a street corner in Georgetown, Rodney spoke of the crisis of Burnham's rule in Guyana and the need for change:

> The other evening, speaking at another site, I had to draw the analogy, to say what if there ever was such a thing called the Midas touch, which was the touch that made everything turn into gold, then we will have a new creation in this society—the Burnham touch where everything he touches turns to shit. (Prolonged laughter) . . . Let us make it clear that we are not asking persons to enter because of support for our ideology. We are asking for an active

> effort of the people for national reconstruction and national unity on the basis of common sense, patriotism, decency and honesty. We are committed to building a Guyana for the future of the Guyanese children. (Applause)[73]

On June 13, 1980, Walter Rodney died tragically in an explosion sparked by a car bomb. According to a 2016 report on the circumstances surrounding his death, Rodney was a victim of state violence.[74] That same year, Burnham was quoted as saying, "Comrades, we are now in the Roman amphitheater. The lion and the gladiator cannot both survive; one must die, and we know that the People's National Congress will live."[75] Burnham's death came early in 1985, but his tight grip carried on through the aftermath of the seawalls.

A COMPLICATED SENSE OF PLACE

Behind a section of eroding concrete with the date 1964 engraved on the side, a man from the Demerara region recalls when the seawall was first installed. He was a young boy. He expressed frustration while brushing his hand against the numbers. "You see most of these old projects they were doing is damaged." The damage he speaks of is not the kind that occurs over time. He is alluding to the corruption of the Burnham years. "More money for big building is more money in their pocket . . . ," he continued. "These people don't know what they're doing." Deep distrust in politics and hardened skepticism about all things political characterize the mood in Guyana today. This skepticism, perhaps justified in its intensity, extends to the politicized discourses of coastal management in a context in which maintaining seawalls has come at the cost of economic stagnancy and dependency on foreign financing.

Real Guyana, a community blog launched in 2013, laments the government's failure to attend to even the most basic needs of the growing population, including the maintenance of seawalls. Soon after forming a social media presence, the group posted a photographic essay inviting readers to take a virtual tour along the seawall at Georgetown. Images of crumbling concrete, overflowing garbage containers, and a rotting oceanside bandstand were captioned with narratives recalling fond memories of the seawall, "where people congregate to socialize and enjoy the North East Trade Winds blowing in from the Atlantic Ocean."[76] Now, gathering at the seawall invites feelings of frustration and a sense of corruption. "After decades of neglect, the Sea Walls have started to deteriorate at an accelerated pace within the last 5 to 10 years. One by one, every single public space meant to be enjoyed by the public free of charge, has been allowed to rot and crumble."[77] The tour continues, raising questions and sarcasm in response to the government's priorities.

Mr. Lewis, a sharp and uncompromising man of eighty, worked on repairing the seawalls in Georgetown in his prime and helped fix the drainage sluice in his hometown of Buxton, which many years later is still in working condition. Having lived through colonial times, Mr. Lewis remembers vividly a period when the seawalls were built to last. "Everything that the old people put down here lasts for *years*," he tells me. And he is not alone in his feelings.

I took a bus to Buxton to meet Mr. Lewis on a Sunday morning. Buxton has a reputation for social unrest, a site notorious for racial tension and violence. In a country where most villages have optimistic names, like Triumph and Hope, Buxton is known for class warfare and an embroiled and prolonged dispute between the village, property owners, and a racially divided government. During the pre-independence period, the more dominant East Indian population of Buxton was forcibly exiled from the area in an uprising. Many families took refuge in nearby villages. And not long ago, many more fled the area when it was taken over by a band of prison escapees. Mr. Lewis and his family remained.

As I made my way along the street, I heard sounds of singing from a wooden church freshly painted white. I entered the house through a gate leading to a garden with a sour cherry tree and followed a path to steps painted a bright shade of blue. Mr. Lewis arrived on a bicycle and was playfully bantering with his grandson, who was in the kitchen preparing the meat and vegetables for the evening meal. We sat down to talk, and he apologized for not having the energy to take me to the seawall; he had already put in a full morning of gardening and had biked all over town.

As we got comfortable in the living room, Mr. Lewis was quick to comment on the disrepair of Guyana's seawalls. He made a point to emphasize that coastal engineers never think to consult the people who have lived in these areas, who have experienced the struggles and hardships of coastal stabilization over time. "Even though these new engineers were coming out," he told me, "they think it's best for them to do their own thing. But *we* working *so long* on the seawall. That type of work is my life."

Mr. Lewis was visibly frustrated. It was clear that this was not the first time he had spoken on this topic. He had gotten into many heated arguments with government officials who he felt were doing nothing to address the changing conditions of the sea defenses. "I went to them to talk about the sluice," he explained. "I told them that the sluice is too low. . . . They said, 'What do you know about that?' I said, 'Let me tell you, I was born and bred here.'" So he took the matter further. "We had a very heated argument over people's land here. They sent notices to sell the land. I went to the meeting to tell them, 'You should be able to help us get the drainage or do something, but you can't take people's land.'" Despite this, the government went into Buxton and repaired the sluices themselves, only to make the situation worse. "They made a *big* blunder," he told me. "They don't ask the older folks about this place. The two kokers there, they blocked them up, and those were the two we used to drain this entire village."

Mr. Lewis has started to wonder if things might have been better before Guyana gained its independence, when the British created the conditions for construction of the sea defenses. His daughter, sitting nearby, was startled at his comment. Before he could continue, she jumped up and exclaimed, "I am surprised at you, my father! You like PNC government?!" Quick to snap back, he shouted, "None the matter, I don't like *no* government! Me like a government that would deal with the people—it's the *people* I studying."

After a bit of a pause, he collected his thoughts and continued, reflecting on how the problems facing coastal Guyana, including climate change, stem from years of bad governance. "I don't believe it's climate change and those kinds of things," he said in reference to the future fate of the seawalls. "I believe it is *negligence*. . . . I'm gonna tell you this, since we got independence and the Indian interim government [Cheddi Jagan] left this place, we're doomed." Perhaps to make sure that his daughter, who had moved into the kitchen, was still listening, Mr. Lewis shouted, "Because of these people, me does call them *parasites*!" This successfully got an

"*Alright*" from his daughter. He went on quietly, "I does call them parasites, they *living* on you!" He looked back at me and added, "They wicked."

Stuart Hall articulates this complicated sense of place as the "New World" presence, which "is not so much about power, as ground, place, territory. It is the juncture-point where the many cultural tributaries meet, the 'empty' land (the European colonisers emptied it) where strangers from every other part of the globe collided."[78] In this "juncture-point," desires collide and diasporas are negotiated. In Guyana, much of this occurs in the shadow of the seawall, which symbolizes the future of the country in its most fundamental form, as land above water. As Hall contends, diaspora is not meant literally but rather metaphorically, defined, "not by essence or purity, but by the recognition of a necessary heterogeneity and diversity; by a conception of 'identity' which lives with and through, not despite, difference; by hybridity."[79]

CONCLUSION

The story of Guyana relates to many coastal zones that have been left with the ruins of an unjust past, shaped in contexts of racial segregation and white supremacy. Colonial land reclamation projects have sowed the seeds for a lifetime of coastal injustice. Yet the lasting harm imparted by Guyana's colonization has never been fully acknowledged. Well into the 1950s, the Dutch coastal engineer Johan van Veen, along with many others, clung to the belief that land reclamation was a pathway to paradise.[80] Despite the failing conditions of seawalls in the former colonies, coastal reclamation continued to be framed as a solution to poverty. These processes, many still ongoing under the guise of adaptation, contribute to coastal injustice.

As Guyana's vast seawall assemblage intervenes in the flow of time, it also invites unexpected encounters and unforeseen avenues of change. Waves of colonization, experimentation, political turmoil, and disrepair have translated into a situation in which 90 percent of the population lives precariously on the edge of the sea. In this context, placekeeping demands a double confrontation: first, with the residues of historical injustice; and second, with the sentiments of colonial nostalgia that the wall conjures in its gradual decline. Furthermore, as sea levels rise, the seawall's meanings and functions change. The next chapter explores contemporary Guyana and placekeeping efforts that are now being negotiated alongside histories of oppression that underlie competing values of coastal protection, from the desire for more concrete seawalls to the dream of a future protected by a self-sustaining green seawall.

3 Lost Origins

DREAMS OF A GREEN SEAWALL

Along Guyana's fortified coastline are symbols and advertisements painted over a patchwork of weathering concrete seawalls. The messages come and go, either fading into oblivion or replaced by the latest mural or government-sponsored message. One of the more visible sections is stenciled with children's eyes to warn the Guyanese that they are being watched by the next generation.[1] Inadvertently, the eyes also give expression to the hidden fears and struggles of life behind seawalls and the shifting meaning of coastal protection in times of abandonment and disrepair.

When I visited in 2012, a section of Georgetown's seawall had recently been painted over with large bold green letters that read, "MANGROVES PROTECT US FROM THE SEA, LET US PROTECT THEM." The message was created by the Guyana Mangrove Restoration Project (GMRP), which had recently formed to bring awareness to the role that mangroves play in sea defense and sustainable development. In contrast to building more seawalls, the GMRP advocates for the regeneration of mangroves as a natural form of sea defense. Its members, many of whom are women, are building what they call "the green seawall" of Guyana that has historically protected the coast, referring to the mangrove belt that lined the shore before European colonization.

The dream of a regenerated mangrove belt along the coast is a strong one with both real and symbolic manifestations, but it is not shared by everyone. Efforts to maintain the baselines of the past challenge longstanding attachments to place. Concrete is not easily given up. There is a deep connection to the colonial landscape, and efforts to preserve the natural environment entail both a desire for something new and a nostalgia for the past. These tensions illuminate systems of value and conflicting aesthetic desires for permanence. For some, place attachment was formed in concrete. For others, place is a living, breathing landscape that needs nurturing. When histories of oppression underlie conditions of vulnerability, there will always be contradictions in the keeping of place.

Guyana's shore remains entangled in the residues of historical injustice, contemporary global inequality, and a changing climate. Nonetheless, staying in place is still strongly preferred to retreating inland. Within the everyday practice of holding the line are remnants of colonial power dredged up in competing practices of coastal protection and conflicts between "expert" and local knowledge. As such, coastal protection entails politics that question and define cultural, aesthetic, and environmental values. And while the Guyanese are resourceful in finding their own unique ways to hold back the inevitable march of time, their solutions are necessarily ephemeral.

Colonial ruination combined with the burden of maintenance has given rise to regenerative placekeeping efforts on the coast. The story unfolds in the context of desperate times—when rising tides threaten to submerge entire communities and mobile attachments are used to temporarily raise the height of existing seawalls. With climate change, the urgency of maintenance takes on new meaning. However, similar to many frontline coastal communities, sea change is not always the main culprit of trouble—at least not yet.[2] The emerging preference for green coastal protection, particularly the regeneration of mangroves in Guyana, should be understood as more than a climate adaptation pathway but as a placekeeping effort that entails attachments that are not easily relinquished. Maintaining the coast is an act that is negotiated alongside desires for self-determination, the recovery of the past, gendered relationships of coastal engineering, and new and sudden discoveries that have the potential to unleash a very different, and modern, future.

THE BURDEN OF MAINTENANCE

The conditions of vulnerability imposed on coastal Guyana during colonization have translated into a burden of maintenance that is not fully driven by climate change. As projects of development and globalization further exploit nations formerly under colonial rule, maintaining the coastal belt of Guyana involves a new set of players. This includes a new generation of coastal engineers tasked with the impossible feat of keeping the shore in place in contexts of global inequality and within economic systems that are inherently unequal. As a result, coastal engineers must negotiate their strategies in contexts of disruption and through systems of inequity, drawing on whatever limited resources they can access.

Engineers in Guyana face a number of unique challenges that make even basic sea defense very costly. In addition to the deep and ever-shifting alluvial mud that piles along the country's foreshore, the tide changes every six hours, meaning there are two high tides a day. Thus, in order to build concrete seawalls, engineers have to first build reinforcements to keep the construction area dry, which can cost more than the wall itself. Maintaining the engineered coastline of Guyana must therefore be done creatively and through piecemeal efforts. This leads to compromises and interventions that coalesce into what Alejandro De Coss-Corzo terms "patchworks of repair."[3]

In areas where concrete seawalls are prioritized, such as in the capital city of Georgetown, maintenance entails attaching mobile units to the top of existing sections of the seawall to prevent overtopping. Starting in 2013, the Ministry of Public Works resorted to using temporary attachments to raise the height of sections of the seawall in Georgetown where overtopping was becoming a problem. These toppers, a stopgap measure to protect land, property, and infrastructure, were affixed to sections of the concrete seawall to minimize spillage. In this case, the need to raise the height of the seawall was attributed to erosion and shifting mudflats rather than a change in sea level. Coastal engineers observed a change in the length of the mudflat in Georgetown, which they believe is migrating east along the coast.[4] As the mudflat shifts, it affects wave action, rendering the intended design of the seawall ineffective.

With increasingly limited financial resources, the Sea and River Defense Division of Guyana has had to find alternative ways to maintain and repair sections of the seawall. Over the past two decades, riprap, or fragmented rock linings, have become the preferred course of action for dealing with breaches. Most of the materials for constructing riprap walls, including sand and rocks, are less costly than concrete and readily available in Guyana. With price as one of the main considerations for coastal defense, the lower cost of riprap has driven the shift away from concrete. The European Union has provided funding for riprap structures in Guyana, and I was told that this support was instrumental in preventing the Sea and River Defense Division from being decommissioned.

Riprap consists of layers of rock aggregates, like a water filter. Geotextile fabric, a key ingredient of riprap structures, acts as a base layer that reduces the number of rocks needed to stabilize the structure. Geotextile cuts costs by reducing the number of layers necessary. It is the only building material that must be imported. Not surprisingly, it comes from the Netherlands. Engineers have had trouble securing exposed geotextile from farmers and construction workers, who have been known to remove pieces, usually during construction or in the middle of the night. As it turns out, geotextile works exceptionally well as a tarp for transporting rice, sand, brick, and other dusty materials.

As of 2013, 90 percent of the ministry's work was being done in the form of riprap. The repair of concrete seawalls was limited and prioritized only for expanding or developing townships. Section by section, aging concrete seawalls were being replaced or piled over with rocks from the quarry. Guyana's contemporary urban waterfront is now almost entirely reinforced with riprap structures. As an engineer from the Sea and River Defense Division explains, "If you go along the coastline, you will mainly see rock structures. Even if you see the reinforced concrete structure, if you go to the front of that reinforced concrete structure you will see a rock-based protection." One of the benefits of this method is that riprap is a flexible structure, able to absorb wave energy without cracking.

The construction of riprap structures is a global phenomenon and can be found everywhere at the edge of urban waterfronts lining creeks, rivers, and coastlines and homogenizing waterfronts into jagged and uninviting edges. In Guyana, riprap is a relatively new and unwelcome intervention resulting in a coastline that is visually and functionally different, a problem that in many cases overshadows its lower costs and more flexible designs. There is nothing particularly glamorous about riprap, in part because it transforms the edge of the sea into a mound of unwalkable rubble. While the design and placement of riprap is largely informed by engineering principles, the finished structures appear to be unfinished, like a pile of forgotten construction materials.

In contrast to the concrete walls of the past, public perceptions of riprap in Guyana are less than positive. When I spoke to those who live near seawalls, the preference for concrete became clear. A council member in Leguan, an island in the Essequibo River east of Georgetown, draws on a lifetime of experience to observe how the older grouted concrete walls lasted for more than sixty to seventy years. "With this riprap, I'm not so certain," he concludes, estimating that it might last a mere fifteen years. For many, the loose rocks look more like a mound of corruption and squandered money. Furthermore, riprap brings with it new hazards. For example, trash and broken bottles lodge in the cracks between the rocks, cutting children's feet as they play and damaging fishermen's boats at high tide.

I asked one engineer who has been very active in riprap construction along the coast how people respond. He was honest, acknowledging that communities prefer concrete seawalls. Even in his own community, where he replaced an aging concrete seawall with riprap, he was met with frustration and perplexity. "Everybody knows I work in sea defense," he explained, "and they keep asking me, 'When are you going to put concrete between these rocks?' Because they see loose rocks and say, 'You're not gonna pump concrete?!'" The wall he replaced was over a hundred years old and was nonfunctional and had eroded the landscape. He did his best to explain to the community that riprap is "as simple as nature" and that in time it would trap sediment and create a little beach. This was not enough. To appease the unhappy crowds, he agreed to build a little concrete wall on top in areas where people had lost their footpath.

REIMAGINING COASTAL PROTECTION

With the decline of Guyana's first seawalls and the growing contempt for riprap, attention is now turning to nature-based solutions and the regeneration of mangrove forests. In Guyana, there are three species of mangroves. The most common along the coast is *Avicennia germinans*, or black mangrove. *A. germinans* is migratory, building up land for decades at a time before shifting and leaving areas exposed to erosion. These mangroves are sometimes called walking trees in reference to their roots that rise above the water. Rachel Carson, known for *Silent Spring*, studied the edge of the sea, giving voice to the "strange and beautiful" world of coastal mangroves.[5] In the slow rhythms of shore time, Carson saw a process of "continuous creation" animated by "the relentless drive of life."[6] She wrote of the shore as a difficult and marginal place, bound together by alliances of mangroves, fish, turtles, crustaceans, and birds. "Only the most hardy and adaptable can survive in a region so mutable," she said.[7]

Mangroves are both fixed and ephemeral, never fully permanent but slow enough in their movements to breed life and a sense of place. The slow mobility of mangroves has also found its way into the reframing of human cultures. The "timeless meandering" and "world-making" of mangroves is eloquently examined in *Swamplife*, an environmental ethnography by Laura Ogden.[8] Following the metaphor of the mangrove, Ogden chronicles the interconnected creation of human and nonhuman lifeways in the mobile landscape of the Florida Everglades. Elizabeth Rush similarly employs the metaphor of rhizomatic flux to illustrate the sentience of coastal ebb and flow in her journalistic account of managed retreat in the United States.[9]

Over time, Guyana's coastline has experienced a steady degeneration of mangrove forests, with a sharp decline in the past few decades. From 1990 to 2009, mangrove forest coverage along the coast dropped 75 percent.[10] This trend parallels global declines. An estimated 35 percent of the world's mangroves has been cleared in the past century, with 50 percent of this the direct result of aquaculture and shrimp farming.[11] Seawalls and other hard coastal defense structures have played a particularly destructive role in the disappearance of mangrove forests along Guyana's foreshore, interfering with the natural cycle of mangrove regeneration and migration.[12] Yet the process is not easily noticed and therefore is often attributed to other forces. Some argue that the trend was set off by economic hardship in the 1970s, which drove people to use the mangroves for fuel and firewood. Other culprits include the use of bark for leather tanning and brick manufacturing, as well as clearing mangroves for fishing access, development, and, most recently, aquaculture. Some of the decline is also due in part to natural erosion.

In response to the loss of mangrove forests, the GMRP formed in 2010, funded by the European Union. Annette Arjoon-Martins, a prominent Amerindian Guyanese environmentalist and pilot, was asked to chair the organization.[13] During her many flights over the coast, she noticed a decline in mangrove coverage and saw that these same areas were becoming more vulnerable to erosion. In a presentation to representatives from the European Union, Arjoon-Martins projected an aerial image of Guyana's receding shoreline. The scene, captured by Arjoon-Martins herself, shows Dutch sluices 170 feet beyond the seawall. These partially submerged structures are the remains of floodgates installed during the construction of the original colonial drainage system, parts of which have been relinquished to the sea.

The GMRP started as a three-year pilot program with the intention of receiving additional funding as well as implementing the first round of mangrove restoration experiments on Guyana's muddy foreshore. Regenerating mangroves as a natural form of sea defense is one of the driving forces of the GMRP. I observed the daily activities of the GMRP during its third year as it transitioned into the Mangrove Restoration and Management Department within Guyana's National Agricultural Research and Extension Institute. During that time, the GMRP staff included a community outreach coordinator, an administrative assistant, a handful of mangrove forest rangers, and a cross-village network of seedling growers and planters. The main site of GMRP activities, including community gatherings, took place east of Georgetown in Victoria, a village celebrated as the first plantation purchased by formerly enslaved persons.

The process of mangrove restoration in Guyana, from harvesting and planting to monitoring, is multifaceted and requires a diversity of knowledge and labor at each stage. Mangrove restoration draws from the knowledge of women, farmers, drainage and irrigation workers, coastal engineers, and scientists who are attuned to the unique and intimate conditions of life on the shoreline. Mangrove restoration is also part of an international conversation that is largely shaped by privileged actors that can claim "expert" knowledge.

The process of mangrove restoration is divided into three stages. First, mangrove seedling growers are commissioned and trained by GMRP staff. The work of harvesting is largely done by single mothers from rural villages. They are paid a fixed rate for each seedling that reaches a height of two feet. Many of the women are new to growing but have thrived nonetheless. A single grower can produce as many as six thousand juvenile mangroves. As of 2012, forty-eight mangrove nurseries produced 210,000 seedlings for coastal restoration.

The next stage involves the planting of mangrove seedlings in the muddy foreshore. Juvenile mangroves are planted along sites where conditions are expected to be favorable for mangrove regeneration. This relies on a mixture of local knowledge and scientific projection, a process that is by no means an exact science.[14] The planting itself is a difficult and physically demanding task and done mostly by men: the mud is deep enough to consume a human. Drainage workers are employed for the laborious work of planting at low tide. To move along the mud, they must scoot on wooden planks and paddle with their hands, much like a surfer riding a board out to sea. The spectacle often attracts a small crowd and a few puzzled looks. Tens of thousands of seedlings are planted in this way in a given area.

Finally, the newly planted areas are monitored to prevent tampering. In Guyana, landless farmers are one of the biggest challenges for the GMRP. Cash-crop farmers use mangrove branches to grow their yields, and landless farmers in Guyana release their cattle to graze along the coast. As farmers leave their goats and cattle on the roadside and the seashore, the animals graze indiscriminately, even wandering into sites where mangrove seedlings have just been planted. The newly planted seedlings are especially enticing for the animals, goats in particular.

The GMRP's dream of building a green seawall is not limited to the more developed regions of the country. It is a dream that extends across the entire coast. In November 2012, I accompanied a small team from the community and monitoring division of the GMRP on a mission to explore the possibilities of expanding regeneration efforts into areas west of the Essequibo River. A mangrove specialist was also present, bringing technical assistance to the project. Starting in Georgetown, we began the overnight journey by crossing the Demerara River on a floating toll bridge built in the 1970s. As we entered the west Demerara region, we arranged to cross the much larger Essequibo River on a small motorboat. The boat felt as if it would burst into a million pieces with each thud against the river rapids but got us safely to the Pomeroon-Supenaam region of the country.

The Essequibo region is primarily a farming area with large rice crops. When seawater breaches the sea defenses, it is very difficult if not impossible to produce rice again. We set out to meet with the local authorities and the Neighborhood Democratic Councils (NDCs) to discuss possible planting sites where conditions of accretion and erosion are ideal for mangrove restoration. The regional executive officer at Anna Regina informed us that the mudflats, which have been gradually building up for the past few years and serve as a form of protection against the sea, have been causing flooding during periods of heavy rainfall. They block the channels, flooding the low-lying areas nearby, while also creating a barrier between the waves at high tide and the land. In the officer's words, "It protects one way, but floods the other way," a problem that is "more good than bad" because breaching seawater poisons the soil.

At Leguan, we gathered at the NDC office in a building seemingly frozen in time, with features such typewriters and large handwritten logbooks. The council members were older and referred to each other as comrades, brothers, and sisters. The conversation started with the challenges that the community faces with regard to mangroves, including the lack of a patrol program or authority to apprehend people to prevent mangrove cutting. They also commented on the declining population of the island, reporting that Leguan went from a population of around eight thousand to four thousand in recent years. All of these factors contribute to the lack of awareness of the role that mangroves play in protecting the coastline from erosion.

The vice chair emphasized the importance of involving people in the work of coastal protection. The chair of the council added, "We were not aware of the importance of the mangrove—no one was—because we were not anticipating the rising of the tide." Now they are more aware of the importance of mangroves and the dangers of removing them. He asked about the feasibility of growing mangroves in an unstable sea, noting that "the riprap is very expensive." He referenced what he had seen while surveying the island in his boat. In line with the conversation, the mangrove specialist added a notion that echoed a larger trend in coastal protection among the international community: "We need to work with nature, not against it."

The GMRP draws inspiration from international conservation initiatives, including a recent shift to community-based conservation. The shift has accompanied a change in ideas about the role of people in conservation, particularly for women. While many of the women employed by the GMRP fill traditionally feminized roles, the program has challenged gender dynamics, giving women agency to work in roles that best fit their needs. For instance, women are employed as mangrove rangers, a role that has been historically reserved for men in forestry. "Not every female would like the idea of being in the field and walking in mud," one of the female rangers told me. "Some days you're wet head to toe." Yet she feels that this is also the best part of her job because it allows her to spend time with her children on the weekends. "They are more excited to go than me," she laughs.

Historically, women have been depicted as either victims of environmental degradation or benefactors of environmental programs. Initially left out of state development programs and international financial institutions, women are now viewed as "indispensable actors in the quest to eradicate poverty."[15] Jill Belksy, an environmental sociologist, argues that women do more than simply benefit from these programs; they "creatively think and reshape development and conservation programs to meet their own strategic goals and agendas."[16]

The GMRP is an award-winning model of community-based conservation and in many ways exceeds the expectations of its funders. However, as community-based conservation becomes the gold standard for international funding agencies, these programs are not perfect, and some have introduced new power struggles and uneven development practices into historically contested sites of injustice. Peter Wilshusen writes about the ways in which power, authority, and legitimacy shape the efforts and actions of agents in community-based conservation projects.[17] Those who advocate for the role of communities in resource management often tend to view a community as a small spatial unit, a homogeneous social structure, and a set of shared norms. Rather, communities are better understood as consisting of multiple actors with competing interests that shape and influence decision making.[18]

While mangroves are appealing for their benefits for communities and the environment, much of their funding is driven by cost. Compared to the construction of concrete seawalls, mangroves can be planted at a fraction of the expense and at a rate much lower than riprap defenses.[19] Money spent on seawalls is money that is not going to social services, which are woefully lacking in much of the country. In some areas, where mangrove coverage has been more consistent, there has not been a need for elaborate systems of concrete or riprap seawalls. This has been the case in Liberty and Hog Island, both surrounded by mangroves. If the GMRP is able to successfully bring mangroves back to areas that rely on seawalls, the newly regenerated forests would help extend the life of existing seawalls and save money on maintenance.

For funders, Guyana's mangrove restoration efforts are regarded first and foremost as a coastal protection measure with a lower price tag. These lower costs are attractive to international development funding agencies. For example, the European Union initiated the Global Climate Change Alliance (GCCA) in 2007 to "strengthen dialogue, exchange of experiences and cooperation on climate change with developing countries most vulnerable to climate change."[20] The GCCA funds projects all over the world and has been earmarking funds for mangrove regeneration under their mission to promote soft adaptation measures and so-called sustainable coastal zone management.[21] As a soft engineering alternative, mangrove regeneration thrives in the neoliberal culture of resilience and adaptation planning. Yet similar models have been tested elsewhere with less than favorable results, for example, in the regeneration of terrestrial forests. One study shows that manufactured forests fail to re-create the ecological resilience and biodiversity of the forests they replace.[22]

A SEARCH FOR LOST ORIGINS

While the regeneration of Guyana's mangrove forests can be seen as a mitigation and cost-saving measure, it also entails politics that are complicated by historical injustice. As a community-based placekeeping effort, regeneration entails what Stuart Hall identifies as a "nostalgia for lost origins," enacted through representations of better times.[23] Mangroves are envisioned as a "return to the beginning," a dream that Hall argues "can neither be fulfilled nor requited, and hence is the beginning of the symbolic, of representation, the infinitely renewable source of desire, memory, myth, search, discovery."[24] Most of the time, these unrequitable narratives—for instance, going back to nature and being free from the dependencies and injustices of the modern world—are not apparent to those who tell them.

Emma Marris writes that restoration starts with the construction of a baseline—a scientific before and after—that is often established as a time before Europeans arrived and destroyed the planet.[25] In Guyana, such a baseline is being imagined and tested through the regeneration of mangroves. Here regeneration is intended not only to preserve existing forests in decline but also to resurrect and create new landforms in places that have been made vulnerable by colonial transformation. Mangrove restoration engages with a vision of the coastline before colonization and a dream of transforming eroding shorelines into resilient coastal environments.

Despite the many benefits of mangrove forests and the potential that regeneration holds for liberating Guyana from its postcolonial concrete dependency, it has been difficult to convince the public that mangrove restoration is the best course of action. Mangrove forests are not viewed as desirable landscapes. In Guyana, people refer to mangroves as the "bush" or "courida." One GMRP ranger explained that the average person in Guyana would not recognize the term "mangrove." Furthermore, perceptions of mangroves stem from colonial worldviews. As such, mangroves have a way of mirroring the tensions of conquest and disparity. Even the scientific classification of mangrove species traces the discursive categories of colonial divisions. Mangrove typologies separate species into "Old World" (eastern hemisphere) and "New World" (western hemisphere) taxonomies, based on characteristics of their evolutionary development. Regenerated mangrove areas are often referred to as mangrove "plantations."

The people who reside in the mangrove forests are some of the most marginalized residents of coastal Guyana. They face widespread discrimination and resentment related to their impoverishment and unwillingness to relocate. One engineer I spoke to at the Sea and River Defense Division felt that coastal vulnerability in Guyana had less to do with climate change and more to do with mangrove squatters. He articulates a view that many share, illuminating the stigmatization of poverty and its association with mangroves:

> I am not one of those who jump too much at the whole climate change thing. People are squatting on the embankment, and they shouldn't be squatting on the embankment, they should go into housing and get a plot and go live there. But if you go squat in the embankment, your activity destroys the embankment and creates erosion, and people get flooded and that's it. Can you now say that is climate change? No. It's causing us a headache. . . . So you can't really just sit down one morning and think that everything is climate change when you have basic things that need to be put in place and cannot even be put in place.

In her study of urban waterfronts in Martinique, Mélanie Gidel draws connections between how mangrove settlements and the squatter communities that are often forced to live in these precarious environments have historically been detested.[26] She writes, "Long seen as hostile and useless, mangrove forests have recently been recognized as valuable ecosystems sheltering a number of species and limiting risks of coastal erosion and marine floods."[27] Recently, mangroves have been framed in relation to climate change mitigation, particularly the connection between mangroves and carbon sequestration. Mangroves sequester ten times more carbon than any other forest in Guyana.

To convince the public of the value of mangroves, members of the GMRP worked with the local media to create a public broadcast spotlighting Vreed en Hoop, one of the most notorious squatter settlements in Guyana. Known colloquially as "Plastic City" in reference to the wrinkled plastic tarps that serve as rooftops, Vreed en Hoop is home to an informal settlement hidden in a thicket of mangroves—a perfect example of the power of mangroves to regenerate land. Ironically, Vreed en Hoop translates from Dutch as "Peace and Hope." Life is hardly peaceful or hopeful for the people who live there. Plastic City is a source of contempt for those who have fond memories of the coast, lost to the mangroves and the undesirable externalities of extreme poverty. In just the past decade, desperation has driven hundreds of people, labeled squatters, to take refuge there, many of whom have constructed illegal dwellings ten to twelve feet above sea level with discarded and repurposed materials.

I accompanied the team during the production of the GMRP broadcast. As we entered the mangrove-seized landscape, it was clear that Vreed en Hoop possessed the evidence that the GMRP was looking for. The area, once an artificial beach with a stone and concrete seawall, had been fully repopulated by mangroves, wrapping the five-foot-high seawall and its expansive groin into a blanket of twisted roots under an umbrella of shade. The groin, also called Vreed en Hoop, stretches perpendicularly from the seawall for half a mile. The land around it accreted rapidly in a manner that was largely unwanted and unplanned, transforming the seawall and its surrounding infrastructure into an unintended sidewalk.

Dave Martins, lead singer of the Tradewinds, a classic band popular throughout the Caribbean, lived in the area during his youth.[28] Dave reminisces about "a very simple life. Nothing much to do; no electricity, no running water, not even a radio in the house."[29] He recounted his initial astonishment at seeing such a profound change in the landscape when he visited for *Stabroek News*.[30] "I was on the Stabroek Market wharf, looking across at Vreed en Hoop, and I was shocked," he wrote. "All I could see was mangrove bush: the groyne was no more; the wooden tower at the end, no more; the stone tower in the middle, no more; the concrete structure, gone." He remembers the groin as "a quiet windswept place" with a "strong flavor of solitude."

> What I knew as a youngster was a naked groyne with nothing growing on either side of it. What is there now at the edge of Vreed en Hoop is actually a dense mangrove forest with towering trees, many 20 feet high, holding the old concrete groyne in its belly. You can't see it from afar—the foliage is too dense—but closer in there's this narrow break in the bush, and the groyne, a horizontal backbone, virtually intact, slicing a tunnel through the forest.[31]

A photo from Martins's past shows the light tower and groin surrounded by sea, without a mangrove in sight. As we ventured deeper into the marshy forest, we came upon the light tower. It was daylight, and a small stream of sunlight trickled through the thicket, illuminating a path for small children to ride their bicycles back and forth along the narrow path. Pink birds with long slender beaks rustled through the mangrove roots, searching for shrimp. The scene reminded me of a passage from *The Hungry Tide* by Amitav Ghosh: "The specialty of mangroves is that they do not merely recolonize land; they erase time. Every generation creates its own population of ghosts."[32]

Later, Martins shared with me a storyboard of an unproduced music video that speaks to other changes he has witnessed along Guyana's coast. As I read through the script, I could see that Davė was witnessing Guyana's natural environment succumb to the colonizing concrete of modernity.

BUILDING WITH NATURE

The process of returning Guyana's foreshore to its premodern conditions is necessarily paradoxical, entailing many negotiations with changing natural processes, technological innovations, and the engineering dominance of former colonizers. Regenerating mangroves for coastal protection falls into the emerging design paradigm of "building with nature" that is embraced by engineers and designers who value green infrastructure as a means of development. The paradigm accompanies the environmental turn in coastal engineering science and the principles of soft engineering, specifying a preference for flexible structures made up of a mixture of inorganic and organic materials present in nature that work in harmony with the sea.

In practice, this approach is often pushed by Dutch engineering firms and sought out by city officials and planners as a technical solution to managing postdisaster landscapes in the developed world. It is also increasingly pursued by developers who wish to capitalize on the improvement of impoverished coastal landscapes in the Global South. However, it has been argued that building with nature is largely a rhetorical strategy to promote seaside development and that in practice hard structures continue to dominate so-called green designs.[33] Such efforts could also be described as a form of "coerced resilience," where a desired ecological state is achieved through human intervention that requires continual anthropogenic manipulation.[34]

The shift to building with nature in Guyana emerged informally as a response to the initial failure of mangrove regeneration. Restoration has proved difficult in the shifting mudflats. In Cove and John, one of the first test beds for the GMRP, twenty-eight thousand newly planted seedlings washed away in high tide. This initial disappointment drew attention to the need for a better understanding of the local accretion and erosion cycles. The GMRP called on the University of Guyana for ideas, engagement, and solutions, leading to a plan to bridge nature-based engineering practices with regeneration.

One Guyanese engineer, who represents a new wave of coastal engineering, offered a potential solution. During her studies, she had seen sand-filled geotextile tubes installed in other parts of the world as low-cost breakwaters. She felt confident that such a design would work well to prevent mangrove seedlings from washing away. The GMRP provided her with funds to build a system of geotextile tubes to break the wave energy at a future planting site in Victoria. This was an experiment. Geotextile tubes had never been tried on the alluvial mudflats of Guyana.

Early in the morning, near the Mangrove Visitor Center in Victoria, I watched as the newly formed crew went to work implementing Guyana's first pair of geotextile breakwaters. One by one, we climbed over a six-foot seawall and up a ladder onto an enormous dredging pontoon temporarily beached on the muddy foreshore at low tide. From the pontoon I could see the first of the two geotextile tubes. It was arranged parallel to the seawall about fifty feet away. The plan was to pump sand and water into the second tube, which was still deflated. The work had to be done in the small window of time between low tide and high tide. Mangrove seedlings were scheduled to be planted in the area soon, and their shallow roots would need protection from the forceful waves that were sure to crash against the concrete divide at high tide. If successful, the mangroves would return the favor of the breakwater by tangling their roots into the ground and growing to protect the weakened seawall.

The construction equipment was contracted through BK International Inc., Guyana's largest privately owned Indigenous construction operation, boasting the largest pool of assets in Guyana. BK projects on the coast typically consist of repairing seawalls and installing riprap structures. Not surprisingly, there was some tension among the crew concerning the shift to mangrove restoration. One of the BK supervisors often gave the new GMRP engineer a hard time, arguing with her over details, sometimes slowing his efforts and making the job more difficult. Yet there was also an expression of recognition through common struggle. At one point, the BK contractor told me that he can relate. He faced a lot of conflict when his generation of builders changed their method from concrete walls to riprap structures.

On site, there were about twenty men working, both young and old. Some were contracted laborers. Others were skilled workers and project managers. I watched as the higher-ranking workers lounged in the shade. From time to time, the laborers would cool themselves in the pool of water created between the pumping machine and the sand pile. One stayed on the geotube keeping track of its inflation. I was amused by a boy playing near the pump discharge. It reminded me of something children do at water theme parks. I asked the project engineer if she was concerned that curious children from the community would play on it. She said she was, mostly that they would cut it to see what is inside. But, regardless, it would soon be covered with rocks, to protect the material from the sun.

The scene generated curiosity and confusion among community members from the small village of Victoria, one of the many areas in Guyana that has suffered frequent flooding and overtopping. The pump that was being used to draw sand and water into the tube was the same kind of pump used for mining operations in the country's interior. A bystander told me that in all his many years of living on the coast, he had never seen a pump like that on the foreshore. "Are they mining for gold?" he laughed. Another bystander asked, "Why are they placing them parallel to the shore rather than perpendicular to it?"

The process was painstakingly slow, with frequent breakdowns. The second tube was beginning to seem futile after days of contractor issues, pump failures, difficult weather, and uncompromising tides. At one point it was determined that the workers were going to need to drag one of the deflated tubes into a new location. I took video footage of the process to pass the time. From my vantage point, I watched the workers struggle to shift the tube. The scene played out like a Charlie Chaplin film performed on the stage of a slippery mudflat. My footage shows scenes of the workers moving the mat one tug at a time, rolling in the mud, and getting a lift into the pontoon by the claw of a big yellow Hyundai tractor.

Later that week, a high spring tide was expected, and we prepared to witness what was sure to be a spectacle. The unusually high tides would provide a small window of opportunity for the contractor to move the giant barge from the muddy foreshore on which it had made its residence. For days, the heavy barge remained motionless beyond the towering concrete wall at Victoria, sunk in the knee-deep mud like a permanent fixture on the murky horizon. With each passing day, it became increasingly difficult to imagine how such a monstrosity would find enough water to be towed out to sea. We arrived at ten in the morning, and to our surprise and disappointment, the barge was gone. It was moved in the early hours of the morning while we slept, and all that remained were the tracks, creating a shallow but wide river in the mud.

By the end of the week, the tube was starting to show some progress. I listened to words of hope that geotextile breakwaters might not only protect newly planted mangrove seedlings but also provide a low-cost and practical alternative for the difficult road ahead. On the last day of my visit, the breakwater was completed. As the geotube rose from the seabed, the crew celebrated their small victory. Months later, an image of newly planted juvenile mangroves circulated on social media—an early indication of the project's likely success.

Mangrove forests are not just restored; they are *designed*. The assemblage of geotextile tubes imported from the Netherlands and mangrove seedlings grown in Guyana is a perfect example of what Emma Marris calls "designer ecosystems" made possible by "less than authentic shortcuts."[35] In Guyana, these shortcuts dredge up relationships from an unjust past. Restoring a balance to an ecosystem that was disrupted by colonial extraction requires the technologies and actors that previously benefited from colonization. One engineer expressed frustration with this underlying problem: "When a consultant comes to a Third World country, they come with a frame of mind that Guyana doesn't know anything. This is why they're here, why they bring them in, because there's no capacity. But that's not the case! At the end of the day this is our place. We know the ground more than any person that comes from a foreign place."

Part of the problem is that "building with nature" draws on an ethos of Dutch superiority over all matters of coastal vulnerability, turning a blind eye to histories of colonization. Henk Ovink, appointed by the Dutch cabinet as the first special envoy for international water affairs in 2015, claims that the Netherlands is particularly well suited for building with nature because the Dutch approach to coastal transformation evolved through a cultural process spanning many years, based on a desire to make land collaboratively.[36] He emphasizes the Delta Plan as a model, citing community involvement as the mechanism of its success.[37] This is despite the controversy of the project, which amounted to a complete reconfiguration of the Dutch landscape and involved blocking several estuary mouths to reduce the length of the seawalls while keeping open the important shipping routes to the ports of Rotterdam and Antwerp. The dikes along these waterways were heightened and strengthened without parliamentary approval in what has been described as another link in the "technological lock-in" of the Dutch landscape.[38]

While the new generation of coastal engineers in Guyana find some inspiration from Dutch engineers who now espouse nature in their designs, they are keenly aware that solutions for Guyana are not easily understood by those who do not experience both the daily and generational fluctuations of moving mud, changing rivers, and rising seas. Even as the concept of building with nature promotes progressive principles of design, innovation, and creativity over technical problem-solving, the practice is entangled in histories of injustice, where entering into foreign lands and transforming coastal landscapes went unquestioned for hundreds of years.

COASTAL PROTECTION AND ATTACHMENTS TO PLACE

Placekeeping in Guyana is being negotiated at the interface of competing logics of coastal protection—from concrete seawalls and riprap structures to regenerated mangrove forests. Desires for protection come into tension with attachments to place. Like hard structures, mangroves are intended to protect the coastline for human settlement, but they carry very different conceptions of place tied to diverse sentiments, livelihoods, and expectations. While mangroves entail hope for the stability of coastal villages in Guyana, they come at the cost of things that many have come to love: space, breeze, and a view of the sea. As such, regeneration is not readily welcomed by those who live near the seawall in Guyana who fear the consequences of an unruly and "bushy" forest of mangroves taking over the more familiar view of concrete, sand, mud, and waves. Mangroves change the essence of place.

One woman shared with me her thoughts on how mangroves complicate her own sense of place. "Almost all my life I grow up by the sea," she explained. "You could stand on the seawall, and see the waves coming, But not now, just trees." Her memories of the seawall start with sandy beaches, white shells, her father's fishing boat, and lifting stones to catch crabs. The scene was not completely devoid of mangroves; in fact, she recalls studying in the shade of the mangroves as a girl. Over time, the sandy beach gave way to mud, and the mud brought with it more trees. Now a forest of mangroves beyond the seawall makes it impossible to see or access the waterfront. "It was very exciting," she remembers, "but not anymore." An uneasy feeling accompanies the inability to see beyond the mangroves. Her neighbor, a woman in her eighties, cautioned me against going near the seawall. "It's dangerous," she says, "because you know what? The thieves." Not to mention the mosquitoes.

For others, the regeneration of mangroves has become a means to reshape and reimagine the coast to meet their unique needs. In addition to harvesting mangrove seedlings for new growth, the GMRP has encouraged the harvesting of by-products from existing mangrove forests to encourage their protection. The Mangrove Reserve Producers Cooperative Society, consisting largely of single mothers, was organized across several rural villages to offer training in beekeeping and to provide opportunities for mangrove producers to participate in farmer's markets and benefit from Guyana's nascent ecotourism industry. A special Mangrove Reserve label was created for the planters to market their honey, pepper sauce, green seasoning, tamarind balls, and other artisanal goods and crafts.

Women derive real benefits from the bees they keep in the mangrove forests. In fact, the relationship between bees and mangrove producers is said to work in tandem with the aims of mangrove protection, as bees require healthy mangrove blossoms to pollinate and get the nectar and producers require healthy bees to harvest honey.[39] Among the women there is a joke that the mangrove producers are creating a line of "bee defense," since bees safeguard the mangroves that protect people from the sea. I spoke to one of the honey producers in more depth. She expressed her newfound sense of independence. "I like to have my own money," she explained. "I have children, and I like to buy things for my children." She grew up in the hills, away from the shore, and moved to the coast as a young adult. In 2010, as she was harvesting honey in the black mangrove forest, she encountered the GMRP. With her help, honey became a staple of the mangrove project.

The intention of the cooperative is to enable women to make their livelihoods from sustainable mangrove harvesting, with the hope that these practices will be passed to future generations. This hope emerges from the failure of the government to protect coastal communities, prompting some to take matters into their own hands. For the GMRP, this includes enabling women to provide for themselves in a context where men and jobs have been largely absent and unreliable. As a result, mangrove restoration work in Guyana has evolved into a women's movement to bring greater food security and economic independence to villagers along the coast.

Not surprisingly, mangrove producers have become outspoken advocates for mangrove regeneration. Unlike seawalls, which have historically enabled extractive large-scale farming for export, mangroves present local opportunities for sustenance and income. I sat in on one of the semiannual meetings of the women mangrove producers in the area and asked whether they prefer seawalls or mangroves in their communities. The response was unanimous.

PRODUCER A: Growing of the mangrove, because you can't get nothing from the seawall. We wouldn't get anything from the concrete wall. It's more expensive. You're going to have to maintain it.

PRODUCER B: When you going to cross the Berbice River you will notice all of the mangrove that protect the river, and if the mangrove can protect the river, the mangrove can *protect us from the sea*. So, if we are going to get that, we prefer one hundred times the mangrove, because realistically, we see what is going on.

PRODUCER C: And it's also good for the environment. The mangrove captures the carbon in the environment, so it's also good for environmental purposes.

PRODUCER D: Yes, you can't get anything from the seawall. The wall is rocks. But if you get the trees there, you know what the trees does for climate change.

PRODUCER A: You have to replace it all the time or the salt will wash away the concrete. And it cost a lot of money to build the wall.

PRODUCER D: So you have a lot of things coming from the mangroves that can benefit members from the community. For example, as a result of the mangroves being there, we could have got into beekeeping.

PRODUCER E: The dry mangrove can be used for cooking. The flavor of the wood, you can barbecue, it has a pleasant and out of the world taste.

PRODUCER A: Yes. With the mangrove, we could get so many things from that, especially the honey. You can't get nothing from the seawall, the wall cannot give us anything, but with the mangrove, we can get honey, and honey is money. Because we are going to sell our *honey* and get *money*. [Laughter.]

PRODUCER B: So one thousand times we prefer the mangrove plants!

Caught up in the enthusiasm, I asked if they prefer "mangroves" or "woman-groves." They were quick to respond, "Woman-groves! A woman's grove, not a man's grove, woman's grove!," and the room burst into laughter.

Among GMRP participants, there is also a sentiment of sacredness that resonates with the ecofeminist philosophy of Vandana Shiva.[40] While not a self-described ecofeminist organization, the GMRP stems from the desire to turn gender disparities in Guyana upside down, arguing that women are indeed closer to nature and thus the privileged agents to lead the struggle for environmental justice.[41] As the feminist scholar Carolyn Sachs notes, the association of women with nature is based on a fundamentally dualistic epistemological strand of Western thought that juxtaposes culture/nature, male/female, reason/emotion, and mind/body.[42] Such juxtapositions have served to situate men above women and nature and have been used historically by scientists to justify and legitimate the domination of and control of nature and women's bodies.

The tension between seawalls and mangroves underscores the ideas, norms, and values that shape relationships to the shoreline. As Kum-Kum Bhavnani, John Foran, and Priya Kurian write in their edited volume, *Feminist Futures*, "Struggles over resources in the Third World are shaped not only by material forces and political power, but also by the ideologies and understandings of what is meant by the environment."[43] In Guyana, the environment is a place that is shaped by historical injustice and contemporary crises of political oppression, global inequality, and climate change. Placekeeping is complicated by the struggle to define and maintain a sense of place in contexts of perpetual disrepair and abandonment.

CONTESTED FUTURES

As people witness more frequent breaches along Guyana's coast, many begin to wonder what the future holds. Ms. Marta, an elderly woman who lives near one particularly vulnerable section of the seawall, sees the situation as extremely urgent and worries about the dangers of a breach. The water now comes up to the top of the wall at high tide—a height that exceeds the top of her home. She wonders, "If this wall should break now, what could happen to us?" She worries that it might take her with it. It won't be long, she thinks, maybe a couple of years.

When successful, mangrove forests create work for women but block access to the sea and its breezes. This has already made life along some parts of the coast fundamentally different, generating a feeling of dislocation similar to coastal retreat. Furthermore, while mangroves absorb wave energy, they cannot prevent the rising sea from overtopping the existing walls in the next fifty to one hundred years.

Given the dire situation along the coast, retreat inland to the hinterland would seem to be one possible course of action. However, some public officials are doubtful that coastal retreat is a possibility. In many ways, the interior of Guyana is seen as a dangerous and disease-ridden place, and for some, relocating from the coastline to the interior is a less desirable option than living with the constant threat of flood. The former minister of public infrastructure told me that abandoning the coast and moving to the hinterland is an unlikely scenario and struggled to imagine a future in which it would be desirable. He added, "Unless we are forced to or unless a tsunami comes and wipes us out, we wouldn't want to do it, because the bulk of livelihood and agriculture comes from this water, and the greatest investment in infrastructure and water management is on the coast."

The heavily forested interior of Guyana is a vastly different and unfamiliar place. Wilson Harris, a Guyanese poet and novelist known for giving expression to the dynamic landscape of the hinterland, remembers his first journey in 1942, when he went into the forests of Guyana as a land surveyor for the government. He described it as "another planet, a living, unpredictable planet," in contrast to the "land one had assumed to be insentient in the coastal and urban regions."[44] While Guyana's interior remains a mystery to the outside world, the reach of the plantation has extended inland through the familiar model of exploitation. As Rodney observes, "Village labor had laid the groundwork for the economic exploration of the hinterland, and this ranks as a decisive and lasting contribution to the political economy of Guyana in modern times."[45] Scattered mining expeditions have brought short-term migrant laborers from the coast into the interior where Indigenous Amerindians live.

Retreat inland is viewed as a development opportunity by some. I spoke to an engineer from the Ministry of Public Works who believes that sea levels are rising and considers it unsustainable to continue to concentrate development on the coast. "We could lose everything in one tide; it's alarming," he told me. He believes that retreat inland could spell a better future for Guyana, although he realizes that it is a slow process. He suggests that it requires the construction of a road, around which development can flourish.

> It can't happen overnight. You can't just tell the people they have got to move, [that] we're losing this war. You just can't do that. You'd get a riot. . . . Bit by bit, eventually we have to move there, maybe another fifty years. It's eventually inevitable, we have to go to the highlands. And that's the rich land, forget about the farming land! All the minerals. We have oil, we have manganese, we have copper, uranium, everything in the mainland.

While some argue that developing the interior is inevitable, and indeed, mining activities are on the rise, one must stop and ask: How would a scenario of coastal retreat affect the forest and the Amerindians? The interior of Guyana remains a nearly impenetrable and undeveloped forest where many of the remaining Amerindians reside. Internal relocation in Guyana is implicated in deeply unjust colonial divisions that extend beyond the coast. As Shanya Cordis, a Guyanese American anthropologist, writes, "The specific experiences of Indigenous peoples in the hinterland and those of Afro- and Indo-Guyanese on the coastland are fundamentally linked. We must disrupt spatial binaries to pinpoint how extractive forces cut across racial hierarchies of power and difference."[46]

Despite the many dilemmas inherent to the maintenance of seawalls, most Guyanese are not eager to abandon the struggle and move to the interior. For many, the fertile coast still has fruits to bear, no matter how difficult they are to nurture. For Ms. Marta, the coast is her identity, a living memory of the past, and a place where she hopes to remain.

> People don't want to move, unless it's a disaster. They see a disaster, and then they would move . . . because they planting rice, they planting sugar, they planting everything from the coast. Nobody wants to go inland. But they say in a couple of years' time we got to go down. I don't know. I don't know. . . . It all depends. Because, let me tell you something. Your parents died here, and that is some part of something that you don't want to move, because every time [you] pass and see this place, there is something within them that [you] want to keep alive.

A civil engineer from Essequibo, who lives half an hour away from the seawall on higher ground, explained that he can sleep better at night knowing that his house won't flood. He has tried to convince his parents to move slightly inland, but they will not until it is absolutely necessary. For him, sea defense is political, and he believes that no one will admit that what is being done is not adequate to safeguard the country. I can see that he has thought a lot about this. Before I left Essequibo, he gave me a sheet of paper with a diagram and scribbled notes on it. He had brainstormed what he thought the social effects of rising sea levels might be for the people of Essequibo. This included rising costs of living and more destructive flooding.

As coastal protection is being reimagined through placekeeping efforts like the regeneration of mangroves, another recent development now threatens to overshadow the work of the GMRP. In 2015, ExxonMobil located a large reserve of oil and gas offshore of Guyana, generating a lot of excitement and controversy about the future of the country. The sudden transition of Guyana from an agrarian society to a petrostate is an ongoing development that mirrors colonial relationships of extraction. Cordis writes that "oil mimics a much older extractive frontier logic that birthed the modern Caribbean and drove the search for gold, the theft of Indigenous lands and territories, and the reduction of enslaved Africans to fungible property."[47] The discovery of oil also has implications for existing placekeeping efforts to protect coastal areas. Environmentalists have raised concerns over how to mitigate the nonhuman environmental impacts of the oil and gas industry, with new criticisms brewing over the implications of climate justice and the future of coastal Guyana in the context of sea change.

The environmental justice implications of hazardous waste and the ecological impacts that a spill would have on critical infrastructure as well as the livelihoods of fishermen is one major concern. These are already exploited environments and people. Arjoon-Martins, with the GMRP, has been a vocal advocate for environmental protection, calling for more oversight and an environmental fund to increase disaster preparedness. Without an official national contingency plan in place, oil and gas activity is bound to have an adverse impact on marine wildlife. While initial environmental impact assessments indicate that the water temperature would make an oil spill less impactful in Guyana, Arjoon-Martins argues that context matters. She notes that there are no baseline studies: "You cannot, in my opinion, manage something that you have not measured. So, if you are starting with these very contentious baselines, we are already starting off on a very weak foundation for monitoring what we have."[48] In an approach similar to her previous work in turtle conservation and mangrove regeneration, Arjoon-Martins wants the communities surrounding Shell Beach to be trained in oil response and to act as first responders.

The problem is that the management of oil reserves is fraught with corruption. Jan Mangal, former presidential adviser on petroleum, cautions that oil is being managed in a "den of corruption."[49] Mangal left his position in 2018 to work independently to, in his words, "perform unpaid advocacy for the country of my birth to ensure Guyana's people and future generations benefited fairly from the country's natural resources."[50] Mangal warns that the financial benefits will not trickle into Guyana or translate into infrastructural development if there are no systems in place to prevent corruption.[51] His focus is on maximizing the revenue for the country and building institutional capacity and anticorruption measures. He argues that accountability requires an "empowered people" to take these companies to task, something that he believes is missing in Guyana—a sense of outrage over the injustice of foreign development and local corruption. "This is not the fifties and forties with the British and sugar. We need to behave like a nation. We need to behave like we *own* this nation. But we're not doing that right now."[52]

Beyond the possibility of an oil spill, oil extraction is bound to exacerbate climate change, a problem that is already threatening the coast of Guyana. This concern has been raised by a newly formed people's movement under the banner, "A Fair Deal for Guyana / A Fair Deal for the Planet." In its campaign manifesto, a connection is made between global climate change and the contradictions of drilling for oil. It reads, "The seas surrounding Guyana are under threat by big oil companies. We believe that the Government of Guyana is putting the natural world and our common home in danger."[53] Citing the experiences of countries like Nigeria, Angola, Chad, and Venezuela, the manifesto emphasizes how oil extraction leads to corruption, environmental injustice, and economic dependency while contributing to climate injustice.

> Meanwhile, no one seems to be paying attention to the impact of oil production on global climate change—the biggest threat to life on earth. So far, ExxonMobil has estimated over 3.2 billion barrels of oil in their concession. Once used, this will release over a billion tonnes of CO_2 into the atmosphere. Guyana will go from being a carbon sink, that prides itself on having vast areas of virgin rainforests, to a carbon emitter.[54]

Guyana's climate future is now at a crossroads, with one path leading to a future of carbon mitigation and sequestration through mangrove regeneration and another path leading to oil extraction and the transition of Guyana to a petrostate. In this context, coastal adaptation is an afterthought, obscured by the semantics of national progress. Yet in the face of erosion, overtopping, and competing logics of coastal protection, the desire to stay in place is alive and well.

CONCLUSION

As the map of the world bends to the anthropogenic forces of sea change, the artificial shoreline of Guyana lives each day on borrowed time. In the seawall's decline are the residues of colonization that have placed Guyana at a disadvantage for maintaining its own shoreline. As is the case in all inhabited areas reclaimed from the sea, particularly those that were taken and altered during periods of colonization, Guyana's coast is one of perpetual struggle—a fragment of the Holocene compelled to survive by suspending its form in time. Responses to the decay of the landscape entail both a desire for self-determination and a nostalgia for colonial values that maintained order along the coastline. These tensions are also implicated in a deeper and more widely experienced struggle for identity and belonging—a struggle that is never fully complete. Meanwhile, the vitality of the intertidal zone yields many transformations that push back at the stone and concrete of coastal colonization, silently reclaiming its stolen pieces. As Wilson Harris observes, "The landscape is alive. . . . [I]t is a living text."[55]

The dream of a green seawall conjures long-standing desires for coastal perseverance. For some, these changes are difficult to embrace. For others, they bring opportunities for regeneration and a chance to return the coast to a former version of itself. This complicated sense of place, caught in the injustices of the past, takes us to the next chapter, where seawall *futures* are being negotiated alongside the threat of rising sea levels and authoritarian development. Across the ocean, another seawall story is unfolding in the small island nation of the Maldives. In the Maldives, placekeeping similarly engages with oppression and contested desires, including a struggle for democracy and small island justice.

4 The Great Wall of Malé

In the Maldivian capital of Malé, one of the most densely populated islands in the world, a giant seawall wraps the outer edge of the coral reef, encircling the island and its inhabitants behind a formidable wall of concrete. As described in the short film *Wave of Change—Maldives*, the seawall "rises from the Indian Ocean like a ringed fortress."[1] The narrator even goes so far as to call it "the Great Wall of Malé"—a foreboding reference that conveys the drama of the seawall that encompasses the tiny Maldivian island, a nod to the technological euphoria of modernization, and the troubled relationships brewing within. Many see the Maldives as a paradise replete with pristine beaches and exclusive luxury resorts, free from the pressures of modern life. However, Maldivians see their world quite differently. They see their islands being rapidly consumed by concrete.

The Maldivian photographer and storyteller Hani Amir documents an intimate portrait of life behind the seawall in Malé in a series of haunting images captioned with his own reflections. Next to a monochromatic photograph of a woman looking out over the seawall, he writes, "You're trapped on this tiny island with nowhere to go. You run to the edge and you're greeted by the seawall that surrounds the island. Not a beach. Just more concrete."[2] Elsewhere, Amir reflects on the desensitization one feels when faced with a life surrounded by concrete: "You learn to play with the concrete. It smells familiar. It smells comfortable. . . . I wonder what the older generations must think of the concrete monstrosity that Malé has become."[3]

With growing international attention to the anticipated losses of climate change, the Maldives has become the poster child of the climate refugee narrative. As one observer put it, "Journalists, environmentalists, scientists and island inhabitants have all clamoured to speak for low-lying islands and their populations, shaping powerful yet disputed discourses about the 'disappearing islands' that contrast with limited attempts to engage in the governing of these island states."[4] This narrative is effective in drawing attention to the distributional injustice of climate change, but the motivation for showcasing frontline communities as the victims of fossil fuel greed rests on a romantic depiction of Indigenous, premodern societies threatened by faraway industrial practices.

The reality is that climate vulnerability and the need for seawalls in the Maldives is contextualized by a recent process of rapid development that emerged through a politics of authoritarianism. Modernization has transformed the country, erasing tradition and thousands of years of sustainable livelihoods, introducing new social and spatial inequalities as well as unanticipated coastal vulnerabilities. Therefore, it is not climate change alone that threatens the Maldives; the project of urban development has undermined the natural defenses of the country and is the foundation on which contemporary placekeeping occurs.

The Maldives has long been characterized by monarchy, centralized power, and political oppression, but it was not until the mid-twentieth century that political elites began to reinforce their power through infrastructural projects that introduced uneven development and spatial inequality to the Maldivian landscape. Since then, the country has become deeply divided by roads, land reclamation projects, artificial islands, and a tourism industry that claims entire atolls—all of which involve seawalls. Modern methods of coastal protection have enabled processes of land grabbing in the Maldives, serving the objectives of those in power while materializing configurations of spatial inequality. Uneven development is both a local and a global process tied to the logics of global capitalism.[5] As infrastructures of expansion and fortification, concrete seawalls are similarly guided by these logics.[6]

Seawalls have accompanied the centralization of resources and power, as well as the expansion of land reclamation for the construction of artificial islands. Over time, seawalls have stretched the outer limits of the Maldivian coast and destabilized complex ecosystems that have long protected the rural livelihoods of Maldivian people. Their ascribed purpose has shifted with the changing priorities of modernization and development. By the time Japan entered the picture in the 1980s, the seawall in Malé was associated with land reclamation and the expansion of development in the context of land scarcity and coastal hazards. Now, with increasing awareness of sea level rise, seawalls have become synonymous with climate adaptation.

The centralizing and authoritarian configuration of power that undergirds the construction of seawalls in the Maldives is hidden by a discourse of disaster mitigation and climate adaptation. Sea change has incentivized urban development and plans to consolidate Maldivians into a future where urban artificial islands encircled by seawalls are offered as the only viable and imaginable solution to climate hazards. Kasia Paprocki conceives of this phenomenon as one of *anticipatory ruination*, where those in power, including experts from abroad, make use of climate discourses for the sake of urban expansion and development.[7] In the Maldives, tsunami recovery and mitigation efforts, combined with the future threat of climate change, are being used to greenwash unjust and unsustainable development plans under the ruse of adaptation.

A BRIEF HISTORY OF UNEVEN DEVELOPMENT

Until the 1950s, the Maldives remained largely undeveloped. Despite British colonial presence from the late nineteenth century into the early twentieth century, the Maldivian people sustained a rural and subsistence-based existence across the many inhabited islands of the expansive archipelago.[8] In 1954, Philip Crowe, US ambassador to Ceylon (renamed Sri Lanka in 1972), traveled to the Maldives aboard a British Royal Air Force Sunderland. He was amused to find that the information network for the entire country consisted of one small printing press, built sometime in the nineteenth century, and two radio operators whose only apparent work was to send regular weather reports to trading vessels. Two decades had already passed since Hoover Dam had been constructed in the American West.

Infrastructural modernization arrived in the Maldives in the early 1950s, initiated by Mohamed Amin Didi, the Maldives' first elected president.[9] While in power, Amin Didi constructed modern roads across many of the inhabited islands of the Maldives, starting with Malé. Enchanted by the broad and open style of the Champs-Élysées of Paris, Amin Didi sought to transform the Maldivian landscape into the image of modernity, with expansive boulevards and plazas. He instructed his engineers to cut roads through the center of inhabited islands, drawing a line from one shore to the other. The process was documented by Crowe, who took particular note of the strong reaction of the Maldivian people.

> It was not until the president embarked on road-building on a wide scale that he earned the real hatred of the populace. As can well be imagined, there is not the slightest reason to build roads on tropical islands, most of which are less than a mile long and half a mile wide. Shady paths between the coconut and breadfruit trees are all that the people want or need. . . . The island looked as if a giant knife had sliced a clean gash across its middle. On this island alone [Malé], I was told, more than five hundred coconut and breadfruit trees had been cut down. So furious were the people at this useless avenue that they never used it if they could help it.[10]

These same roads led to Amin Didi's undoing. His presidency was overturned after less than a year, and he was banished to internal exile in Kaafu Atoll, where his health quickly deteriorated.

The legacy and memory of Amin Didi underscores contentious notions of progress during this period. Some Maldivians remember him as a hero. In "A Tribute to the Late [Amin] Didi," the propagandist poet Ali Shareef praises him for his accomplishments, including "ending isolation" and "the days of successive rulers."[11] However, there are other accounts of Amin Didi that are not so celebratory. These focus on his corruption, including spending government money on houses for his wives and girlfriends while the Maldives was in turmoil. Regardless of how Amin Didi is remembered, his roads remain in place today and have provided a pathway for modern development.

By the 1970s, the pace of development picked up to such an extent that the British anthropologist Clarence Mahoney concluded, "There are few places in the world that have experienced outside influence as rapidly as the Maldives did in one decade, during the 70s."[12] The model of development was adopted straight out of the textbook of the US-based modernization theory that was its contemporary. Modernization theory gave development and urbanization planners the justification and moral authority to industrialize former colonies, based on the assumption that it was "necessary to liberate Third World economies from traditionalistic restrictions to rational dynamism."[13] Perhaps the Maldives' saving grace was that it was too far from the Cold War to be anyone's pawn and the state budget too small for the country to fall into the clutches of the International Monetary Fund and the World Bank.

From 1978 to 2008, Maumoon Abdul Gayoom monopolized political power in the Maldives, leading the country into a period of rapid development and urbanization. Gayoom styled himself president, head of the judiciary, and the highest religious authority in the country, "winning" six elections in a row for the Maldivian People's Party without an opposition candidate. As the *Economist* colorfully put it, he was "an autocratic moderniser who made the Maldives the wealthiest corner of South Asia by promoting high-end bikini-and-booze tourism (usually on atolls some distance away from the solidly Muslim local population)."[14]

Gayoom modernized the Maldives through land reclamation projects, coastal armoring, and the construction of an artificial island, reinforcing spatial inequality at each and every turn. Pictures of Malé from the early 1970s show lush green trees on both sides of the controversial road that intersects the island. In just a couple of decades, Gayoom grew Malé into the most densely populated island in the world, with an imposing skyline of buildings, very few green spaces, and a complete lack of natural beaches. Limited employment opportunities and the concentration of resources in Malé gave rise to internal migration and flows of circular migration in the Maldives, causing people to relocate to the city. These decisions were particularly difficult for families, who were often separated for long periods of time.

The decision to fortify Malé with a massive seawall has come at the expense of the country's smaller inhabited islands. By concentrating development in the capital island, including the majority of governmental spending on education and health, Gayoom reinforced class division and contributed to a growing rift between rural and urban livelihoods.

Some believe that Gayoom's plan to modernize the country was an intentional scheme to concentrate power and development in Malé. The former minister of environment, Mohamed Aslam, explained in an interview, "If Gayoom had foreseen the problems that would arise with economic development and the infrastructure that comes with it, we would not be in the position that we are in now. I think that he didn't want to have other parts of the Maldives developed." Many of the difficulties that small islands now face exist in relation to the overdevelopment of Malé.

Centralized development in Malé has translated into underdevelopment for the Maldives' small inhabited islands, introducing relationships of dependency to formerly self-sufficient communities. For example, many of the smaller islands now rely on desalination plants in Malé for their primary source of drinking water. As one government official explained to me in an interview, the promise of Malé's modern desalination plant inadvertently compromised the entire country's relationship to water—the primary source of life and subsistence for islanders: "Traditionally, every household had a rainwater collection facility, but now they have got rid of these rainwater harvesting systems and have gone for bottled water. Changing rain patterns makes it unreliable. Water is desalinated in Malé, and that has to be shipped to the islands. That's the only form of water security." In 2012 alone, he received seventy urgent complaints from the nearly two hundred inhabited islands reporting a lack of water.

Access to clean water is further compromised by the implementation of poorly designed, low-cost government projects throughout the country's smaller inhabited islands. One woman who lives in Thinadhoo, a populated island in the north, complains of PVC pipes bursting open into the road, water contamination, and flooding: "In the road near my house, there are two sewage pipes that go to the sea. Salt water is getting into the junction via the pipes, blocking the pipes. If I am not home, my home gets flooded." Freshwater exists in a thin layer below the land and above the seawater, making it particularly susceptible to saltwater contamination and pollution from sewer lines.

Gayoom's modernization agenda also led to the expansion of the tourism industry, which has created further inequities in the Maldives. In Benjamin Sovacool's study of expert views of climate change adaptation in the Maldives, one respondent reflected on how economic polarity maps onto the country's scattered island geography: "The Maldives is really two blocks of islands, those that act as tourist resorts and are populated almost entirely by tourists and resort staff, and those home to ordinary Maldivians. The first group of rich islands can afford things like harbors and land reclamation, the second group cannot."[15]

A behind-the-scenes tour of Bandos Resort reveals the hidden infrastructural privileges of tourism and the modern amenities that enable luxury islands to exist. At Bandos Resort, machines operate behind carefully camouflaged buildings that blend into rows of bungalows scattered throughout the island. Bandos was built in 1972 on an island near Malé by the family of a former Maldivian prince. Like a stationary cruise ship, the island is equipped with modern luxuries, including its own supply of desalinated water and imported food as well as a system of interisland transportation and sociopolitical autonomy. The manager of Bandos Resort took me to the far end of the island where the desalination plant is located. With a hint of technological enthusiasm, he showed me the spinning dials and knobs indicating the amount of power being generated from the deafening machines behind the wall.[16] On nearby islands, people struggle to find basic access to clean water.

As is the case in most small island nations, the benefits of tourism are not felt by ordinary Maldivians. In the Thinadhoo region, for example, there are several five-star resorts in close proximity, yet the people living there do not benefit from the jobs created. As one woman told me, "There isn't any tourism work in this area. . . . Our knowledge doesn't work in the tourism industry." Many of the resort workers are migrants from Bangladesh who live on-site sometimes for years at a time. Furthermore, waste regularly finds its way to neighboring islands. A small farmer from Thindadhoo explained that "the beach is filled with garbage including the reef. This is a huge problem. We can't even get into the water from that area." A woman from the same community said that she has to keep the doors closed due to the flies.

A government official told me that even a modest increase in tourism taxes would enable Maldivians to help themselves. A more substantive raise, he argues, would "provide a lot of social infrastructure for roads, sewage, housing, harbors, and island protection." However, the stranglehold of power prevents serious consideration of this possibility.

There is little talk of taxing and regulating the tourism industry. As he explained further, "We have businesspeople who own resorts, and they have counterparts in the parliament, so it's difficult to get anything through with this. . . . There was so much lobbying not to have any taxes on these things." Without government regulation, tourism continues to exacerbate spatial inequality and land scarcity in a country that prioritizes infrastructural enclaves for the wealthy and the powerful.

JAPANESE SEAWALLS IN THE MALDIVES

The overdevelopment of Malé would not have been possible without outside support. The construction of seawalls and other economic development projects were enabled by the technologies and funding mechanisms of foreign investors and international aid organizations in the spirit of neocolonialism. On April 11, 1987, Malé was flooded by a storm surge, displacing hundreds of people while destroying much of the island. The disaster prompted a plan to construct a mega seawall around Malé. Funded by Japan's International Cooperation Agency at a cost of $60 million, the project took over a decade to complete. At that time, Japan was a rising superpower looking to enhance its economic and geopolitical influence through foreign aid and infrastructural development projects.

Modeled on the modern coasts of Japan, the seawall project in Malé included the use of massive tetrapod units cast in concrete. Unlike previous designs that made use of boulders and rocks, tetrapod units interlock to create a porous barrier and rarely become dislodged from one another over their estimated seventy-year lifetime. The construction of Malé's seawall follows a global wave of coastal transformation in which tetrapod structures were being used to convert shorelines into concretized landscapes—a phenomenon that has been described by the designer and architect Bret Milligan as a "sublime corporate geomorphology experiment."[17] Invented in France in 1950, four-legged tetrapods were widely adapted all over the world. As of 1996, when *History and Heritage of Coastal Engineering* was published, it was estimated that more than three hundred tetrapod structures had been constructed around the world.[18] Later designs, inspired by the tetrapod, included more complex shapes, such as dolos (developed in South Africa in 1963), antifer grooved cubes (developed in France in 1973), and hollow or multi-hole units (developed in the United Kingdom in 1982).[19]

I caught up with Takaaki Uda, one of the coastal engineers who worked on the tetrapod seawall in Malé during the 1980s and 1990s. It had been nearly thirty years since the project's commencement, and he was delighted to revisit the subject. He described Malé as looking "like a Tokyo," in reference to the island's out-of-place skyscrapers. His office was lined with bookshelves containing a career's worth of reports. At the small conference table in the center of the room, diagrams were neatly piled showing the contours of Malé before and after the tetrapod seawall was built. "This is the island, the original shape," he explained, pointing to the edge of the coral reef. "Strong waves northeast and southwest can break due to the very wide coral reef." Then, over time, as the city grew, humans became situated very close to the edge of the reef, damaging it and narrowing its width.

Uda explained that when he came to survey Malé in 1987, the Gayoom administration had already been experimenting with expanding the size of the island by dumping sand in shallow areas and holding the land in place with a small seawall. These efforts were the real cause of erosion on the island, and the construction of a tetrapod seawall was viewed as an opportunity to double the size of Malé. However, this would come at the expense of the island's coral reefs and natural defenses. "But I understand," Uda told me, "this island should be used for human beings, *only* for human beings." For him, the decision was clear that the island must be protected by using a seawall. He went to work measuring the island topography in preparation for what would be the largest development project in the Maldives.[20]

The synergist relationship between the Japanese government and the Gayoom administration was facilitated by a shared logic of modernization and economic consolidation. In Japan, the political centralization of coastal defense started in the mid-twentieth century. Kiyoshi Horikawa dates the origins of coastal engineering in Japan to 1953, when Typhoon Tess struck Ise Bay, killing nearly four hundred people and inundating half a million homes.[21] The disaster prompted the creation of the Sea Coast Act, shifting responsibility to the governing bodies of coastal prefectures while standardizing the design and execution of coastal structures to protect against tsunamis and storm surges. Before the Sea Coast Act, seawalls and other coastal defense infrastructures were constructed independently and in piecemeal fashion, prompted mainly by domestic trade along the coast. Countermeasures in the early twentieth century largely involved relocating homes to higher ground, with few exceptions.[22]

By the mid-twentieth century, as the field of coastal engineering matured and industrial expansion swept the country, the Japanese began to increasingly develop hard structures in combination with tsunami evacuation warning systems. Through a process of trial and error, the Japanese modified the types of seawalls used to protect shorelines but based their heights on the crests of the most recent and devastating tsunami. During this time, much of Japan's coast became lined with concrete tetrapod structures. Stephen Hesse, a writer who lives in Tokyo, describes the divergent ways in which people see and describe the concrete fixations: "Massive and practical, cute and pretty, irritating and ordinary, queer, sexy, perverse, sophisticated, a blight on Japan's beauty."[23] One company even marketed plush tetrapod throw pillows that "turn your couch into a coastline."[24] As Sarah Baird, a science writer at Ars Technica observes, "Tetrapods have become such an ingrained part of the Japanese landscape that they've even gained some footing as a pop culture touchstone."[25]

In Japan, seawalls are known to engender a false sense of security, creating imaginary geographies of risk that materialize into unintended consequences.[26] At Tohoku University, I met with one of the leading experts on tsunamis in Japan who has been working on modeling disasters for over five decades. He shared with me a hazard map of Ōtsuchi—one of the disaster zones—prior to the 2011 tsunami that devastated the northern Pacific coast, killing over sixteen thousand people and igniting the Fukushima meltdown. The excessive loss of life in the Ōtsuchi area was due in large part to an illusion of safety. Many of those on the "safe" side of the hazard line did not evacuate after feeling the earth shake, defying the generational wisdom of living in tsunami territory. They felt that the raised seawall would protect them.

The Ōtsuchi hazard map relied on representing the damage caused by a tsunami in 1933, the largest in recent recorded history, and this map was distributed to residents before the 2011 tsunami that shattered records and surpassed the expectations of engineers and tsunami scientists. In fact, government officials and engineers would not accept responsibility for the failure of the physical walls, since there was no precedent for such an event. In many ways, Japan's modern relationship to the sea is stuck in a sliver of time. The shortsightedness of extreme coastal measures, and the false sense of security they create, washes over the longue durée of geologic time and what it means to live with a degree of uncertainty.

While discussing the history of Malé's seawall with Takaaki Uda in Tokyo, I could see that Japan was undergoing its own intense debates concerning seawalls. The failure of both the physical seawalls and the virtual hazard lines to protect people was still very raw. In the wake of the disaster, many preferred institutional approaches, including educational programs, advanced warning systems, and redevelopment on higher ground. The government, however, proceeded with its plan to spend $9.8 billion to raise the height of the physical seawalls another fifteen meters, despite the protests of many scientists, environmentalists, and fishermen. For some areas, this would mean rebuilding walls higher and wider than ever before and in some cases as high as a three-story building, requiring windows to look out to the sea.

Less about preservation of life, Japan's seawalls can be seen as a safety measure for other infrastructural choices, including the proliferation of nuclear power plants, coastal ports, and highways. The height of the wall, determined by the worst-case scenario, can be thought of as a political statement, a manifestation of a national security culture where rare and extreme occurrences are used to justify public expenditure. Perhaps because of this, some Japanese coastal engineers have changed their tune, seeing the error of introducing hard structures to delicate island ecosystems like the Maldives. Looking back at the Maldives and other small island countries, Uda said, "It must be stopped, not only in foreign countries, but in Japan as well. It's nonsense." Yet uneven development is not easily stopped; the seawall in Malé has already put new desires into motion.

THE EXPANSION OF MALÉ TO THE NEW FRONTIERS OF COASTAL DEVELOPMENT

The centralized and massive economic growth that occurred under Gayoom resulted in the overpopulation of Malé. By 1985, Gayoom had expanded the size of the city through the Malé Land Reclamation Project, adding land to the island's outer edges by dredging sand from the foreshore. The project created more space for development and construction, causing Malé to become overcrowded with modern public buildings and facilities. With the construction of the seawall in Malé, the island quickly exceeded its own capacity, and the demand for additional land prompted a plan to expand Malé into a nearby shallow lagoon.

In 1997, Gayoom signed off on the development of an artificial island called Hulhumalé (New Malé), the largest land reclamation project in the country's history. At an initial cost of $32 million, it took five years of dredging to form the base of the new island.[27] Hulhumalé was initially promoted as a project that would address land scarcity and create more space for housing in Malé. As is the case elsewhere in the world, artificial islands are sometimes promoted to meet the civic needs of spatial restriction and city expansion. However, such promises are often lost in the wave of capital investment and used simply as a rhetorical strategy for making development sound less exclusive.[28] This was certainly the case for Hulhumalé.

When I visited Hulhumalé in 2013, it was still in its early phases of development and looked like a ghost town. The reclaimed area, with open and undeveloped land that suggested temporary abandonment, was visible from Malé. The juxtaposition of the old and the new Malé was striking, one island bursting at its seams with concrete and the other an open, dusty blanket taking on the unnatural shape of a rectangle with right-angled edges. Both were suspended in different temporalities of construction.[29] As time went on, non-native greenery was planted in the sands of Hulhumalé in an attempt to raise the environmental appeal and retail value of the land, ironically causing the artificial island of Hulhumalé to appear more natural than Malé.[30]

The first phase of the project was completed in 2004 and was intended to house up to sixty thousand people, with 30 percent of residential development designated as social housing. In 2013, the population was somewhere around fifteen thousand. This was due in part to extravagant housing prices set by the Hulhumalé Development Corporation, out of reach to most Maldivians. One Malé resident told me that he could not afford a house on Hulhumalé despite earning a salary above the Malé standard, explaining that he would have to pay a substantial down payment and holding fee while also paying to live in his existing home as he waited for years to move in. For him, this was impossible. In addition to the high prices of regular units, the subsidized units reserved for social housing were quickly bought up illicitly and rented to the highest bidder. The government did nothing to stop this and instead took two hundred social housing units for itself to auction off in 2018.[31]

It is often assumed that land scarcity is inherent to the ecology of small island nations, where space is so visibly limited. However, few realize the impact that land grabbing has on spatial inequality—a reality that is easily overlooked in favor of building new land through expensive and environmentally destructive reclamation projects. For example, tourism has claimed over one hundred islands as citizens struggle for space. Islands are often bought and sold on the open market to businesses and to individuals who plan to use them for their own purposes.[32] The Maldives is a classic example of what Jamaica Kincaid describes in *A Small Place*, in which she speaks of the tangled web of injustice that tourism engenders.[33] As on many small island countries, high-end tourism has sucked away resources, taken over beaches, created excess waste, and claimed entire islands.

As Maldivians struggle with housing and urban gentrification, the outside world sees artificial islands like Hulhumalé as panaceas and symbols of resilience. Land reclamation is romanticized, often without considering the social or environmental contexts of such projects. With the rising awareness of climate change and the adoption of a climate adaptation discourse, artificial islands are often imagined as a solution for small island communities. For example, when the BBC broadcasted the short film *Wave of Change—Maldives*, created by Journeyman Pictures in 2005, Hulhumalé was presented as an "extraordinary solution."

> The squeeze was reaching a crisis point when President Gayoom came up with an extraordinary solution. And this is his vision: rising out of the sea is the new Malé, Hulhumalé, entirely man-made and the highest, driest island in the entire country. It was to be the foundation of the president's dream of building a new Singapore in the Indian Ocean. Malé's residents jumped at the opportunity to leave a crowded capital behind for President Gayoom's Promised Land a mere 10-minute boat ride away.[34]

Similarly, NPR's Jon Hamilton ran a story in 2008 titled "Maldives Builds Barriers to Global Warming," and in it he argues that "countries struggling with climate change could learn a lot" from the Maldives.[35]

Despite the buzz of international praise, Hulhumalé was not designed to account for sea level rise. Constructed at a height of only two meters above sea level, the island's baseline was determined by the height of the 1987 flood, not predictions about long-term changes in sea level.[36] By 2011, the island was already experiencing drainage problems and showing signs of vulnerability to heavy and severe storms.[37] Nonetheless, the neoliberal dream of endless growth and development continues. By the second phase of the project, Hulhumalé was being called "City of Hope."

GAYOOM'S SAFER ISLANDS DEVELOPMENT PROGRAM

Under the Sixth National Development Plan, Gayoom introduced the National Population Consolidation Strategy, a relocation program to consolidate the entire population of the Maldives into a handful of islands fortified by seawalls. Initiated in conjunction with his development plan, Gayoom's Safer Islands Development Program called for the internal relocation of the Maldivian population into a set of islands modeled on Malé and Hulhumalé, enhanced by means of land reclamation and concrete seawalls. The initial plan anticipated ten to fourteen islands to be engineered as "fortified bases" to withstand sea change while "providing an outlet for development away from Malé."[38] As a hard engineering approach, the plan relied on "the use of human-built infrastructure, such as sea walls and desalination plants, to cope with rising sea levels and saltwater intrusion into freshwater supplies."[39]

The primary justification for population consolidation rests less on safety and more on economies of scale. The plan is guided by the assumption that developing a small number of islands, as opposed to providing basic services across the wide spectrum of populated islands (nearly two hundred of them), is more cost-effective.[40] The cost of producing an acceptable standard of living for the population, including services such as education, health, and infrastructure, is estimated to drop significantly. As the argument goes, building seawalls for every inhabited island is "particularly expensive, leading to one of the highest coastal protection cost-to-GDP ratios of any nation."[41] This sentiment was captured by a former engineer for the Ministry of Public Works, who was sensitive to the fact that population consolidation looks different from a social perspective.

> From an engineering perspective it's something that *has* to be done in a small country because the population is too scattered and there's too much money going on providing basic facilities, and the costs in most instances are prohibitive. Just for effective development to hold, I think population consolidation is the answer. That's from a development perspective, not from a social perspective.

Given the fact that the Maldives spends 30 percent of its national budget on embankments and revetments, it is understandable why Gayoom would aim for economic consolidation. Yet mega seawalls and large-scale land reclamation projects are central to the population consolidation plan, both of which require continual and long-term maintenance.

In 2013, I discussed population consolidation with a telecommunications executive. As he ordered an espresso and two cigarettes, he pointed to the seawall and reminded me that we were sitting in a reclaimed area. Population consolidation would reduce the need for installing and maintaining a growing network of towers on islands throughout the Maldives. Centralized urban development would allow for the company to set up fewer towers, spend less on maintenance, and hire fewer people. Most importantly, the company would not have to cover as much ground in a country without an affordable transport system.[42] "We know from our experience when we built the communication network how difficult it is, even to transport," he told me. "In America you put everything in your van, you go and fix it, but if I have a problem with one of the phone systems, say, in one of the hotels, we have to go by boat." His concern was that bringing modern amenities to small islands, including hospitals, schools, and seawalls, would not be cost-effective. He suggested that population consolidation is necessary, but without a heavy hand from the government, it would not be possible.

Population consolidation has also been justified as the more environmentally responsible pathway of development in the Maldives. One government official with ties to the Environmental Protection Agency in the Maldives agrees. "I think it is very important to consolidate people to the larger islands," he told me. "The government cannot afford to have coastal protection around all these two hundred islands." He argues that this would ease the financial burden of the government while also protecting the environment: "Rather than people living in two hundred islands and destroying the environment, select a few islands and then carry out development on those selected islands; it's much better for the environment."

As the flood of 1987 raised the first alarm about the potential impacts of future sea level rise in the Maldives, Gayoom seized the opportunity to use climate change as a justification to continue with the Safer Island Development Program. During the 1989 Small States Conference on Sea Level Rise, Gayoom welcomed donor countries in a speech stressing the urgency of climate adaptation, concluding with, "There must be a way out. Neither the Maldives nor any small island nation wants to drown. . . . We want to stand up and fight."[43] As one reporter wrote in *Bloomberg Businessweek*, "Gayoom wants more than sympathy. He wants money, namely $1.5 billion in aid to build sea defenses for 50 of the country's 200 inhabited islands."[44]

Forced relocation in the name of urban development is a common experience in the Global South—a struggle around which many social movements have formed. One striking example that alludes to the case of the Maldives can be seen in Franny Armstrong's film, *Drowned Out*, which documents Indigenous farming communities being displaced by a series of mega dams launched in the Narmada Valley of India.[45] The project, embraced by the Indian government as a symbol of national progress, prompted villagers to organize a people's movement. The film documents their decision to stay and "drown" in protest, refusing to succumb to the slower death of displacement and unjust relocation. It follows their hunger strikes, rallies, six-year battle with India's Supreme Court, and the eventual—and inevitable—occupation of their homeland as the water rises and submerges their land. As a result, some of those villagers now find themselves stranded on tiny islands in the vast reservoir behind the dam.

Now, as development discourse shifts from narratives of national progress to climate resilience, development priorities are being reframed and rebranded as *climate adaptation*, a term that provides the perfect foil for contentious relocation schemes. In the Maldives, the relocation of small islanders is promoted as a form of *managed retreat*, defined as the movement of people or property away from coastal hazards. While managed retreat is often viewed as a proactive and necessary form of climate adaptation, it has been criticized for raising significant social and environmental justice concerns and for reinforcing existing inequities and unjust land use practices.

Most studies of managed retreat have focused on the Global North, overlooking the widespread and most insidious forms, including forced relocation and eviction of the urban poor to less desirable locations. Such discussions are often framed within a tradition of Western experiences or a set of misguided assumptions about the Global South and fail to account for the broader set of global issues involved in the production and reproduction of coastal vulnerability across waves of invasion and disruption. In Lagos and Nigeria, for instance, managed retreat is used to justify the relocation of the urban poor from waterfront areas.[46] A similar trajectory has been noted in the Philippines and Colombia, where "land use regulations and evictions—in the name of adaptation" have resulted in the displacement of poor informal communities while allowing wealthier settlements to remain in place.[47] For small islanders in the Maldives, Gayoom's Safer Islands Development Program is a displacement scheme that threatens traditional livelihoods.

Funding support for Gayoom's Safer Islands Development Program came in the wake of the devastating 2004 Indian Ocean earthquake, which reached a magnitude of 9.1 before unleashing a deadly wave of tsunamis. While the tsunami damaged areas of the Maldives, at an estimated loss of $470 million, equal to more than 60 percent of the country's GDP, Malé was largely spared. Elsewhere, the tsunami displaced 7 percent of the population (approximately 29,000 people) and created severe hardship on many islands: 35 percent of the 198 inhabited islands experienced "high or very high impact . . . with major physical damage to buildings, infrastructure, crops and natural vegetation."[48] The tsunami spread waste across the islands, eroded beaches, and damaged freshwater supplies almost everywhere.

At the height of recovery efforts for victims of the tsunami, Malé's seawall made headlines for its perceived role in protecting the capital island. The BBC ran an article titled "Sea Wall 'Saves Maldives Capital,'" putting a portion of the title in quotes to reflect the words of Mohamed Latheef, named a "representative" of the seawall.[49] Lateef advocated for the importance of the wall in protecting Malé and was also quoted as indicating that "he wanted to extend the sea walls to build a 'necklace' around the islands."[50] However, in reality, the seawall did very little to protect Malé. Water went straight through the structure and into the city. Malé was in fact lucky to be in an area that was not as badly hit by the tsunami as other islands in the region. As it turned out, the neighboring airport island of Hulhulé helped soften the blow, and with Hulhumalé still largely undeveloped, there was little to be lost compared to the smaller event of the 1987 flood that devastated Malé.[51]

To take advantage of disaster recovery and reconstruction funds, Gayoom quickly adapted his population consolidation plan to engage with the emerging discourse of climate adaptation, identifying five "safe" islands for the relocation of heavily affected communities. Rural islanders were labeled "vulnerable" and in need of adaptation, particularly those who lived in the devastated island of Kandholhudhoo, one of the largest settlements in the Northern Raa Atoll, with a pre-tsunami population of 3,664. Population consolidation was reframed as "adaptive relocation" and "community reestablishment."[52] The safety rationale was used to deny affected islanders recovery funds of their own. Justifying its actions by sensationalized narratives of climate change and misrepresentations of what islanders actually wanted, the government cut off resources to affected islands, including the possibility of rebuilding infrastructure and providing access to basic services.[53] Kandholhudhoo islanders were left with little choice but to relocate.

Many of the tsunami victims were moved to Dhuvaafaru, a large uninhabited island in the Northern Raa Atoll. Mimi Sheller, who studies post-disaster mobility in the Caribbean, terms this the *islanding effect*, where those in power engage foreign aid to help some of the population while leaving others to fend for themselves—or worse, to undergo forced relocation.[54] Internally displaced communities are further harmed by uneven recovery, in what Sheller describes as "an ongoing process of marginalization, serial displacement, and containment—as if they were marooned on an island of misery, even while surrounded by the coming and going of well-equipped frequent flyers."[55] The relocation process of affected islanders to Dhuvaafaru cast a shadow on Gayoom's Safer Islands Development Program.

To make way for relocation, Dhuvaafaru was stripped and the island topography was regraded. Despite the fact that Dhuvaafaru was without any perceivable advantages in terms of natural protection or elevation, houses and structures were quickly constructed without careful planning, leading to their concentration on the island's more vulnerable outer edge.[56] Perhaps to balance the optics of adaptation, the community center was built on concrete stilts to serve as a shelter in the event of a future tsunami.[57] Conflicts started to surface between the new residents, who were not only from Kandholhudhoo but also from affected islands across several atolls.

In Dhuvaafaru, tensions grew alongside the strain of unemployment and housing inequality, mirroring the injustices of earlier repopulation schemes. For example, in the 1960s, islanders living on the neighboring island to Malé, where the airport is now located, were forced to move to Malé. In an interview, a government official explained to me how uneasy that transition was.

> People might think that because we are a homogeneous society that it would be very easy to move a population from here to there. It doesn't happen like that. There are always conflicts and there is always that line that separates the two communities. The government wanted to depopulate the island to develop the airport. If you ask people here [in Malé] on the road where that community is living, they would be able to point them out. They're still together. How many years has it been since then? I don't know, but so many years, and they are still referred to by the community's name.

Population consolidation raises an important question: Who gets access to protection? According to a UN report on human rights, the allocation of coastal protection is competitive and controversial and without democratic governance, a potential threat to human rights.

> The prioritization of islands for different types of preventive engineering measures, such as protective barriers and reclamation projects, as well as repairs of damage regularly incurred to housing and infrastructures on the islands and environmental projects, is a matter of intense competition and controversy in the country, which must be dealt with in accordance with principles of transparency, consultation and impartial environmental risk and needs assessments. Putting in place such systems is a matter of priority given the growing needs due to environmental pressures, the limited Government budget, and the need to consolidate good governance and the new human rights and democratic institutions in the country.[58]

The use of the tsunami as a catalyst for Gayoom's unpopular population consolidation plan is a classic example of what Naomi Klein terms *disaster capitalism*. In "Blanking the Beach," Klein describes the intersection of the Indian Ocean tsunami disaster with the insidious ambitions of plans to "build back better."[59] As Klein notes, oppressive governments often soften displacement rhetoric, using terms that imply a just outcome while enabling those in power to take advantage of international aid.[60] Uma Kothari, who also writes about the use of environmental discourses and narratives to drive unpopular migration policies in the Maldives, argues that "technocratic and depoliticised discourses of climate change have often been invoked to conceal underlying political agendas in which environmental concerns are drawn upon to justify unfavourable government policies of mobility and resettlement."[61]

The outside world's well-intentioned but insular focus on climate change has obscured the impact that political oppression and development have had on the Maldivian people and their island ecologies, favoring stories that position seawalls and artificial islands against the existential threat of sea change. Disasters like the 1987 flood and the 2004 Indian Ocean tsunami, coupled with the anticipated impacts of climate change, reinforced Gayoom's vision to build more islands like Malé and Hulhumalé, pressuring rural Maldivians to relocate to a handful of designated urban islands.

THE MYTH OF THE "SINKING ISLAND"

Popular perceptions of Maldivians as climate refugees operate through a racialized imaginary of underdevelopment. Such depictions are grounded in what Carolyn Farbotko refers to as "wishful sinking."[62] We (liberal, progressive, Westernized do-gooders) need these images of frontline communities to drive our passions and our convictions forward. But this is not an accurate representation of what is happening throughout the world, nor does it capture what is taking place in the Maldives. Indigenous and Third World peoples have come into contact with development and globalization in various ways and are not passive victims to the external forces of modernization. Those who resist globalization and destructive development are also calling into question their own governments. As Uma Kothari and Alex Arnall found in their research, the urgency discourse of climate change, used among elites, washes over the everyday realities of people in the Maldives.[63]

To make matters even more complicated, it is not clear whether the Maldives is sinking, or at what point the islands' ecosystems might destabilize. Nils-Axel Mörner, who was once a lead reviewer for the Intergovernmental Panel on Climate Change, has been at the center of this debate about sea level rise in the Maldives. Contrary to the commonly held belief that the Maldives would become submerged at even the slightest rise in temperature, Mörner suggested in an article in *Global and Planetary Change* that the Maldives' islands are resilient, having survived historical fluctuations in sea level.[64] Shockingly, Mörner used these findings not only to challenge Maldivian claims but also to furnish climate denialists with a scientific basis for dismissing sea change. In 2007, he was interviewed by Gregory Murphy and claimed that sea level rise is a "total fraud."[65] Mörner later published a rant in the conservative news outlet the *Spectator*, accusing Maldivian politicians of "scare-mongering."[66]

Paul Kench, who spends his professional life studying island geomorphology in the Maldives, as well as archipelagoes in the Pacific Rim, criticized the methodological approach of Mörner's study.[67] Kench has also published findings that challenge assumptions about the future of the Maldivian islands. In one study, Kench finds that land area has actually increased over the past fifty years.[68] In an interview, he explained to me that his intention in publishing the paper was "to inject some sensible understanding of these places." He felt that the sensationalized image of sinking islands was not helpful: "All this talk about drowning and disappearing, that's not a basis on which you can start to plan. It's just a hysteria that whips everyone up into a frenzy." Despite his intentions, the paper was picked up by the conservative media. Kench explained, "They missed the message; it got captured as 'islands growing,' which wasn't entirely true. . . . Then the anti–climate change lobby got hold of it and said, 'Ah, climate change isn't real.' So briefly we were the poster child for the climate skeptics' campaign."

Kench is concerned less with sea level rise than about the ways in which development undermines the natural defenses and life cycles of the dynamic island landforms, rendering them vulnerable to erosion. For him, the one-dimensional climate narrative that is created by well-intended environmentalists undermines the more immediate and addressable issues of maldevelopment. To illustrate his point, he draws on the case of the small island nation of Tuvalu, where journalists travel regularly to take pictures of what appears to be climate-related flooding.

> It's comical that the plane seats about sixty people, and thirty photojournalists get off. The photographers will take photos of this stuff and they say that the islands are sinking. Now, actually, the king tides are the highest tides of the year, and what [you see] is the groundwater table that's popping out of the island's surface because the sea has pushed it up. What they don't tell you is that the island has now become so populated that people have had to move into swamps and really silly places that used to flood anyway.

This trend has also been observed in the small island nation of Kiribati—a phenomenon that Sophie Webber terms "performative vulnerability," where government officials are incentivized to perform their own climate vulnerability in order to secure financing for adaptation in contexts where they have been left out of traditional development funding.[69] In the Maldives, the perception of sea level rise and variants of climate activism are born from these tensions. Theoretical narratives of sinking islands have extended into climate justice claims as well as into legal discussions of state sovereignty and the ramifications of "disappearing islands." Among legal scholars, the question concerns the future of national sovereignty in the case of territory loss and total population migration. As sea levels rise, they ask, at what point will states become extinct?

While the conversation is purely a moral and academic exercise, it gives credence to sensationalized narratives of climate change and is used in the public and by states themselves. Yet the "sinking state" phenomenon, despite drawing much attention to the extreme impacts of climate change, is misleading. Perceptions of environmental migration are often overly generalized and unreflective of local demographic and environmental differences. As the legal scholar Jane McAdam explains, "That is not to say that climate change is not having real impacts on small island States; it is. But the Atlantis-style predictions that have captivated the imagination of some are unlikely to materialise as the means by which States cease to exist."[70] In the Maldives, the anticipation of loss that accompanies climate change is a profound concern, but the destabilization of natural barriers from decades of rapid development and expansion to the outer edges of the reef is more readily felt.

CONCLUSION

The recent physical and social transformation of the Maldives through projects of modernization and land reclamation has created a problem in which spatial inequality and the disruption of natural barriers are as much a culprit of vulnerability as climate change. Contemporary vulnerability to sea level rise, extreme weather, and tsunamis is profoundly shaped by uneven development. And in this context, coastal protection is often narrowly understood as a solution rather than as a driving force of inequity and vulnerability. In the Maldives, the primary function of seawalls is not to protect people from the devastation of tsunamis or rising sea levels. Rather, seawalls are infrastructures of authoritarian consolidation, anchoring centralized resources and economic priorities in politically strategic locations. Seawalls are vehicles of spatial inequality and urban development justified through the logics of consolidation, economies of scale, and, more recently, climate adaptation.

As the Maldives continues to transform through processes of modernization and coastal protection, including the construction of seawalls and land reclamation projects, efforts to consolidate the population into a handful of "safe" islands continues. This is the subject of the next chapter, which focuses on the contest over the future of the Maldives. Placekeeping efforts and the ongoing struggle to define who the Maldives is for and who gets access to protection are being negotiated in a political landscape of competing desires, from authoritarianism to climate justice and a movement for democracy. An alternative to Gayoom's safe island development scheme has emerged: a democratic movement that seeks to bring resilience to diverse island ecosystems without population consolidation and conventional seawalls. However, the effort to move away from hard structures is not always welcomed by those who would stand to benefit from regeneration. This stems in part from histories of injustice that continue to constrain and challenge the prospect of a just and sustainable future in the Maldives.

5 Contested Futures

THE HOPE OF A LIVING SEAWALL

In 2014, I met with a pair of young activists who had recently formed the NGO Save the Beach Maldives.[1] We convened on the sandy shores of the small island of Villingili, a short ferry ride from the capital city of Malé. As I arrived, I realized that I was stepping onto the nearest natural beach for nearly one-third of the Maldivian population who live on an overcrowded concrete island encircled by seawalls. A government official had previously told me that the Maldives "is a country with two or three worlds in it." He was referring to escalating class division in the Maldives, where First World luxuries play out next to Third World scarcities. These worlds include urban islands, tourist resorts for the rich and famous, and hundreds of smaller inhabited islands on which Maldivian culture has taken root over thousands of years.

Thanzy, one of the founders of Save the Beach Maldives, explained to me that Villingili was under constant threat, and as it had the last remaining public beach in the atoll, it was crucial to protect it. About four thousand people visit on a typical Friday, some arriving early to claim space along the beach, which is only a thousand feet in length. As we watched children play in the surf and women enter the water fully clothed from head to feet, Thanzy went on to explain how spatial injustice is felt by ordinary Maldivians.

> It's fine for the government, who will sit in their offices and can probably afford to go to other islands for swimming or picnics, but the truth for the majority of the people in Malé is they can't afford to go elsewhere. Transport is really costly. And the only other picnic island nearby is now closed because the government gave it to a resort, and no resort island would want a picnic island when they can make millions with a resort.

Save the Beach Maldives started in 2008 with the single mission of removing garbage from the beach. Its efforts quickly evolved into a social movement to protect the living shoreline from the forces of modern development, preventing the further construction of concrete seawalls on the few remaining public beaches in the country. "On this beach, right here, they were trying to make a harbor for the police," Thanzy told me. "It would have destroyed the surf. So the surfers had to take a stand against the harbor, and they protested until they decided they weren't going to build a harbor here."

When I came to Villingili, I had climate change and rising sea levels on my mind, but I could sense that this was not the most pressing issue. Save the Beach Maldives was in the process of developing new and more intensive ways to tackle a host of intersecting forces, including erosion, waste, development, government buyouts, and coral bleaching. Sea change was not at the top of the agenda yet. How could it be? The struggles that the young organizers were facing had emerged in the wake of urban development in the Maldives. For them, saving the beach is fixed in the present; without the daily battle on the ground, Villingili and other small islands would not be worth saving in the future.

After spending the day with the organizers of Save the Beach Maldives, I asked how they would describe the mood in the Maldives today. I was met with laughter. Thanzy conveyed a mixture of pessimism and humor: "At the brim of destruction, just on the verge of falling off the edge of the reef." Another response shifted the mood to optimism: "I will say it's turning in a good way, because I want to put some positive energy on it, because we will do our best to stay and help the environment and help the people here."

From my time in the Maldives, it became clear that the multiple worlds of the Maldives are not only spatially and economically segregated, but contingent on who is looking. There is the world that outsiders see, onto which they project their fantasies of paradise and their anxieties about climate change. There is the world that the wealthy and powerful see, onto which they project their desires for more wealth and more power. And then there is the world that the young and the hopeful see, onto which they stage their dreams of a better world. Together, these visions are embattled in competing efforts to survive, shaped by the material realities of development and the politically and ecologically complicated fight for the future. Fueled by a variety of ecological and political pressures, adaptation cannot be understood without attention to the larger story of placekeeping and contested claims over the future.

As seawalls lock the country into cycles of dependency, centralized development threatens to erase the small island cultures that have existed in the Maldives for thousands of years. Place is not only threatened by the anticipation of sea level rise, but also by adaptation schemes that seek to swallow rural livelihoods into urban economies of scale and to dictate a pathway characterized by urban futures and fortified islands. As Paprocki noticed in the case of Bangladesh, "The demise of rural futures is entangled with the celebration of urban climate futures."[2] In the Maldives, the battle for the future is also a fight for the maintenance of rural livelihoods and the contested role that technology should play in adapting to climate change. While some prefer to keep the island ecosystems as natural and as untouched as possible, others dream of fashioning new designer islands out of dying lagoons.

Understanding the dilemma of climate disruption in the Maldives requires attention to how placekeeping operates at the intersection of political oppression and environmental disruption, a struggle that occurs at many scales, from the nation as a whole to individual islands. As adaptation pathways in the Maldives split into opposing political and ideological camps—from more seawalls to relocation to hybrid solutions that mix green and gray infrastructure—it is important to recognize the ways in which infrastructure is being negotiated, reimagined, and reshaped into new ideas for keeping place where it is. Staying in place involves challenging uneven development and an artificial divide between small islands and the urbanized centers of the country, which has implications for who climate adaptation will serve.

REIMAGINING THE MALDIVES

In *The Battle for Paradise*, Naomi Klein writes about two competing visions that formed in the aftermath of Hurricanes Irma and María in Puerto Rico: one that aimed to serve the vested interests of neoliberal elites and another that emerged from the grassroots, a movement for self-reliance and a path forward to justice and sustainability.[3] Similarly, in the Maldives, following the 2004 Indian Ocean earthquake and tsunami, two competing national adaptation strategies emerged: an authoritarian plan to consolidate the population into a handful of "safe" islands encircled in concrete seawalls and a newer democratic vision to usher in a different form of adaptation based on decentralization and soft adaptation measures.

The Safer Islands Development Program—promoted by the Maldives' long-ruling authoritarian leader, Maumoon Abdul Gayoom—prioritizes urban expansion and the development of fortified islands while seeking to relocate rural, Indigenous islanders. Similar to other large-scale development projects around the world, the Safer Islands Development Program embraces a vision of progress that displaces rural livelihoods in favor of urbanization. In a study of climate migration in the Maldives, researchers found a strong preference among islanders to stay in place and "to continue living in their houses with their communities . . . irrespective of potential future climate change impacts."[4] Despite the obvious desire of Maldivians to maintain their traditional livelihoods, Gayoom has pushed for internal relocation and a deliberate shift to modernization.

In the wake of the 2004 Indian Ocean tsunami, Gayoom was pressured to permit free elections in exchange for international disaster aid. During this time, a small window opened for a new democratic movement to challenge the assumptions of centralized development and usher in new possibilities for Maldivians to stay in place without population consolidation. A contrasting vision emerged that included a range of alternatives, from interisland transit to the regeneration of natural defenses. Its advocates envisioned a future in which all islanders can have access to the benefits of development while maintaining their small island livelihoods and fragile island ecosystems—where resilience is not granted by mega seawalls and artificial islands for the benefit of the few.

The young journalist Mohamed Nasheed—previously imprisoned and tortured by the Gayoom administration for protesting human rights abuses—returned from exile to run for president of the Maldives under the banner of his newly cofounded Maldivian Democratic Party (MDP). Nasheed ran with the slogan "Aneh Dhivehi Raajje" (The Other Maldives), evoking the defining slogan of the global justice movement, "Another World Is Possible." In Dhivehi, the native language of the Maldives, there is said to be no word for democracy. It wasn't until the 2008 electoral campaign that a Dhivehi equivalent for the term came into use. Today, the English-language term "democracy" is often used by Nasheed's supporters synonymously with the Maldivian phrase "Aneh Dhivehi Raajje."

Nasheed won the presidency on October 8, 2008, to the shock of Gayoom and his supporters.[5] Immediately, a new development plan was introduced that emphasized progressive change and decentralization. In 2009, with consultation from the United Nations Development Programme (UNDP), Nasheed's government constructed the 528-page "Strategic Action Plan." The vision, which was also articulated in the MDP's campaign manifesto, was written to align with the United Nations Millennium Development Goals and included a plan for ensuring affordable living, equitable housing, and health care for all. The plan also aimed to undo many of the destructive development priorities introduced by the Gayoom administration, highlighting five main pledges, with an explicit intent to ensure social justice "to open up opportunities for the most disadvantaged sections of the society to emerge from their present conditions of poverty thereby helping the country achieve its development goals."[6]

One of the five key pledges was the establishment of a nationwide transport system in response to the tourist monopoly of transportation that has resulted in the physical immobility of much of the population.[7] In the Maldives, transportation injustice and differential mobility is inherent to processes of spatial inequality and uneven development that underlie conditions of vulnerability. Most people cannot afford access to the limited transportation services available. As the plan states, "A sustainable maritime transport network will increase accessibility and mobility of the people and will increase economic regeneration at all levels through revitalization of the urban setting and land use."[8] By prioritizing the mobility of Maldivians, the MDP challenged the assumption that people must be moved to urbanized islands in order to gain access to the benefits of development, evoking the core principles of the emergent concept of *mobility justice*.[9] An affordable interisland transit system would enable people from small islands to access essential services without having to relocate.

The MDP Strategic Action Plan sparked an alternative epistemology rooted in democratic ideals rather than authoritarian mandates—an adaptation pathway that promised development while maintaining the right to remain in place on whichever island one chooses. As Mohamed Aslam, Nasheed's minister of the environment and one of the leading organizers for democracy in the Maldives, told me:

> We need to break free of the idea that this island is too small so we need to move to a bigger island. We need to find a solution *in between*, but there are colleagues of mine who don't accept this. Often, they think in economic terms. . . . All I'm saying is no, you don't have to do that. You don't have to have two hundred gynecologists or two hundred eye specialists or dentists. Find geographically the islands which can be connected via a transport link, provide service for them to wherever island is best suited for people to come and go, and then give them access. We cannot bring down the entire Maldivian population to ten islands or five islands. People can *dream* of doing that, put it in numbers and work out the economics. But it's not simple economics. It's a social issue.

Aslam further explained why population consolidation is a flawed and fundamentally unjust concept.

> I don't think that people are like trees. You can't pluck them and replant them. It's not that simple. We've lived on these islands for so many centuries. We have a sense of belonging to these islands. There's a history. Your memories are attached to that. You feel home on these islands. Your island may be small, but that's where you call home. So people are very reluctant to move to somewhere else. Especially when it is done en masse. I believe there will be people among us who will choose to live in small communities, who will choose to live in isolation, and we will have to accept the fact that that's a right they have, and they should not be deprived of that right. And therefore, we cannot have population consolidation as we had attempted to do in the past.

He repeated the same sentiment at the 2011 United Nations Climate Conference in Durban when he joined forces with civil society activists. Frustrated by the lack of political will, they occupied the conference space and demanded that world leaders take action. Echoed by the "People's Mic," Aslam shouted into the crowd: "We have our rights! We have a right to live! *We have a right for home!*"[10]

Nasheed's development plan envisioned a future radically different from Gayoom's Safer Islands Development Program, entailing a deliberate shift away from the seawalls and hard engineering practices to softer resilience pathways.[11] It was widely recognized that the construction and maintenance of seawalls was siphoning money from public services, contributing to spatial and political polarization while separating those who have access to basic services and coastal protection from those who do not. Embankments in the Maldives are so extremely expensive that the Maldives is already spending over 30 percent of its national income on climate change adaptation. At the time of my research, it was anticipated that adaptation costs would surge up to 50 percent of the national budget. This leaves very little to spend on education, health, housing, and other basic necessities.

NASHEED'S SOFT ADAPTATION PATHWAY

Guided by the ideological frameworks of the international environmental organization 350.org and the British science writer Mark Lynas, Nasheed rewrote the narrative of the "sinking island" to show leadership on climate action. Combining his message of moral authority with a commitment to embracing climate solutions and staying in place, Nasheed turned the focus from fossil fuel restrictions to economic opportunities. This shift marks a deliberate contrast to the more common framing of the Global South as climate victims seeking repayment for a "climate debt" owed them by the rich nations of the world, which developed at the expense of the rest of the world. By the end of Nasheed's first year, the Maldives had become known as the country least responsible for, yet most vulnerable to, the impacts of climate change and, most importantly, the first to take action by pledging to go carbon neutral.[12]

In 2010, Nasheed initiated the Integrating Climate Change Risks (ICCR) program, a four-year, $9.3 million adaptation project funded by the UNDP, the United Nations Framework Convention on Climate Change (UNFCCC) Least Developed Countries Fund, and the Maldivian government. The ICCR marked a discursive shift from Gayoom's hard path of population consolidation to a path of soft engineering and resilient island planning, including the planting of mangroves and coral regeneration as an alternative to more seawalls. The ICCR named "mangrove afforestation and beach nourishment" as an alternative to seawalls and "coral propagation around existing islands" as an alternative to the creation of designer islands like Hulhumalé.

> Instead of using man-made sea walls or tetrapods to counteract rising sea levels, the ICCR programme relies on mangrove afforestation, thickening of coastal vegetation and beach nourishment. Instead of desalinating water to deal with water shortages, it uses larger catchment areas for rainwater and elevated water storage tanks. Instead of reclaiming land to deal with tidal inundation, it uses dune replenishment. And instead of erecting artificial designer islands, the ICCR programme promotes coral propagation around existing islands, all at a fraction of the scale and cost of the SIDP [Safer Islands Development Program].[13]

The ICCR promised physical, institutional, and community resilience through the development of a suite of tools and technologies intended to create risk reduction plans to "safeguard public and private assets worth $20 million through adaptation measures, create at least four Atoll Development Plans, and protect a total population of 42,000."[14] Specifically, it focused on four areas of resilience building: (1) "capacity development," defined as the creation of a climate information system to train government officials how to assess risk, costs, and benefits; (2) "demonstration projects," to test a sample of measures in four locations, including coastal afforestation in Kulhudhuffushi and new coral reefs in Thulusdhoo; (3) "risk reduction," to create disaster risk profiles; and (4) "knowledge management," to integrate Maldivian policy makers into a global platform of experts on climate policy.[15] Implemented by the Maldivian Ministry of Housing, Transport, and Environment, the project was designed to demonstrate the cost benefits of soft measures.

However, there are limitations to these new soft measures, and the ICCR was often at odds with Nasheed's democratic principles. The ICCR is a classic example of top-down adaptation driven by outside influence rather than by community-based planning. As Sovacool gleaned from his interviews with government officials, climate policy remained administratively centralized in Malé.[16] Moreover, the work that Nasheed was doing to decentralize development in the Maldives was complicated by what Adam Grydehoj and Ilan Kelman call "conspicuous sustainability." In exchange for economic opportunity, countries like the Maldives become "trapped as exemplars, struggling to meet potentially unrealizable expectations they have foisted on themselves."[17]

The shift to soft adaptation measures was further contextualized by economic strain. As a poor country, the Maldives is not in a position to build more seawalls. Mohamed Aslam estimated that each meter of coastal protection costs between $4,000 and $5,000. One unnamed government official (and skeptic of the ICCR) argued that raising funds for seawalls in the Maldives is impossible without outside help:

> The sea wall around Malé, for example, cost $54 million to erect, or $12.4 million per kilometer. The Maldives has 2,002 km of coastline, which would make protecting them all with a seawall a monumental $24.8 billion enterprise. With the country's current annual GDP, it would take more than three decades to raise the funds for such a task, let alone build the sea wall. We've also got only $9 million in total to work with for the ICCR. What are we going to do, build half a kilometer of sea wall with the money?[18]

Another government official articulated a more enthusiastic message, equating natural and cost-effective soft measures with Indigenous methods, arguing that policy makers need "to move beyond our bias towards strong, technological and infrastructural solutions."[19] Yet there is doubt that soft measures will be adequate enough to meet the dangers of sea level rise in the Maldives, and these doubts are not unfounded. Experts caution that "adaptation efforts may be insufficient to ever fully eliminate the risks associated with extreme projections of climate change."[20]

THE PENDULUM OF ADAPTATION

The struggle for the future of the Maldives is a contest of competing desires to define and keep place. As politicians vie for power in the Maldives, the pendulum of adaptation swings between democratic and authoritarian principles.[21] This translates into competing and sometimes mutually exclusive visions for the future of the small island nation, from Gayoom's conviction to build more artificial islands to Nasheed's dream of fostering regeneration and enhanced interisland transit. The division is value-driven and revolves around economies of scale, questions of mobility, human rights, and the future of seawalls. The split leaves little room for negotiation, and much like an election, the winner takes all.

When I visited the Maldives in 2013–14, Nasheed was no longer in power and population consolidation was again on the table.[22] The political and economic allies of the long-running Gayoom dictatorship never accepted the results of the 2008 election and through the whole of Nasheed's tenure waged a dirty campaign to regain power. On February 7, 2012, Maldivians were shaken with the news that Nasheed had "resigned" his presidency. Within hours, Nasheed and MDP supporters were seen in the streets of Malé, protesting what they called a coup. They were beaten and arrested by the police and military, by then firmly in the hands of Nasheed's vice president, Dr. Mohamed Waheed Hassan Manik. Waheed proceeded to dismiss the entire cabinet and named a who's who of Nasheed's political opponents to his own cabinet. The coup government turned the page back to the Gayoom dictatorship's repressive politics of beatings, imprisonment, and torture to brutally discourage the repeated street demonstrations in favor of Nasheed.[23]

With the country back in the grip of dictatorship, it was difficult to imagine how Maldivians were going to maintain their vision for "Aneh Dhivehi Raajje" or emerge from the disruption of centralized development. Without Nasheed in power, the Maldives' plans for climate action have taken a backseat. Shortly after Gayoom's return to power, the government announced plans to drill for oil.[24] Shauna Aminath, a member of Nasheed's presidential cabinet, spoke of the irony.

> It is concerning. The people who are "elected," their policy is to find oil in the Maldives! As one of the most vulnerable countries, I don't understand how we can drill for oil. We should be promoting renewables. It is the most bizarre and illogical thing I've heard for the Maldives to drill for oil. . . . Authoritarian values have been promoted, freedom of expression has been constrained, our allies in COP [Conference of the Parties] meetings are now

> distant. . . . We've basically said "no" to the rest of the world. We've asked the world to stay away from the Maldives.[25]

Mark Lynas relayed this point to *Minivan News*: "I think that the Maldives is basically a has-been in international climate circles now. . . . The country is no longer a key player, and is no longer on the invite list to the meetings that matter."[26] The fragility of conspicuous sustainability was made visible as foreign investors pulled out of their commitments to fund carbon neutrality in the Maldives. In a cruel twist of fate, the very afternoon of the coup, the Maldives was set to sign into existence a solar plan that would have launched the country onto a path of carbon neutrality.[27]

As the pendulum continues to swing, and the battle for political power ensues, the anticipation of loss will likely present many more adaptation plans. However, the struggle to stay in place is not only a climate issue. Staying in place is a contest of social and cultural values. Throughout Gayoom's thirty-year dictatorship, many of the unique customs and traditional folklore of the Maldives were criminalized in an effort to impose conservative Islamism on the entire country—a religious identity that many Maldivians consider foreign.[28] When I visited small island communities during my research, a traditional boatbuilder recounted stories of his own customs and traditions, which were outlawed as antireligious.[29] The phasing out of non-Islamic customs has further deepened the divide between rural and urban islands in the Maldives.

After Nasheed was elected in 2008, MDP supporters in Addu city came together to respond to the loss of historical customs by building a cultural center with traditional artifacts, photographs, and oral tradition on display. The center was swiftly destroyed by Nasheed's opponents in a fire the night before the February 2012 coup. During my visit, one of the curators let me in so that I could walk through its charred remains. The curator pointed me to the group's Facebook album, where pictures of Addu Cultural Center were displayed and posted under a caption that spoke to the loss.

> Addu Cultural Center was built with the pride of a beautiful culture and heritage. The project was put together with high spirit and aspirations to share the richness of our profound history and spread the fascinating stories from our past. Exactly one year ago from today (7th Feb), the center was burnt down during the political turmoil. Addu Cultural Center was the first and

> the only such place here in Addu. Many items of immense value were destroyed, burnt down to shards of dust. With it, as a culture and a society we lost irreplaceable treasures of our own past.[30]

These political and cultural tensions are also communicated across conflicting landscapes of artistic representation. After the 2012 coup, the government-backed Islamist Adhaalath Party organized an exhibition at the National Art Gallery, opened by President Waheed (the former vice president who replaced Nasheed). Sixty paintings were displayed under the theme, "Fall of a Regime: An Artist's View," all created by a single artist and painted over the course of just one month. Some of the paintings were direct copies of photographs, with MDP colors and supporters omitted. This unusual effort to illustrate the transfer of power was in all likelihood commissioned by Waheed in response to an earlier announcement by MDP supporters who were independently planning the Exhibition of Public Inquiry (XOPI) at the grounds of the Malé City Council.[31] The theme of this exhibit, "Truth Is Ours," was set to challenge the official narrative by giving space to a wide range of artists to reflect on the events leading up to and following the coup.

One artist told *Minivan News*, "I am participating because this is another venue to express my thoughts and feelings about the coup, freedom, liberty and justice. At a time when our freedoms to assemble and express are getting limited, this space suddenly becomes very important to me."[32] In contrast to the images of "peaceful" anti-Nasheed protests set against the whitewashed walls of the National Art Gallery, an ominous sculpture lingered at the XOPI grounds. The artist wrote a description next to the sculpture: "Grasping to comprehend the reality of the situation and describe something so phantom and menacing in my head was the image of a charging bull at the door."[33] Another painting shows a solitary officer waist-deep in riot gear, suggesting a deluge of injustice.

As rising seas threaten the small island country, one must ask what is already sinking. The result of Gayoom's efforts to depose Nasheed on the basis of fundamentalist religious accusations has escalated into a violent repression of human rights. On August 4, 2014, a young journalist named Rilwan reported on these sensitive matters.[34] Four days later, Rilwan went missing. This generated a massive public outcry on the streets of Malé.

Three months after Rilwan's disappearance, with no answers in sight, discontent was brewing throughout Malé. On the hundredth day of his disappearance, Rilwan's friends, family, and supporters took their protest to the seawall in Malé. Signs were pasted onto the wall's surface showing Rilwan's gentle smile under the bold caption, "ABDUCTED." The seawall was transformed for the day into an altar of hope, despair, and outrage.

THE EMERGENCE OF LIVING SEAWALLS

In the wake of the political turmoil and the impasse of competing pathways of authoritarian and democratic adaptation, informal avenues have opened for young environmentalists to experiment with regeneration. The members of Save the Beach Maldives are engaging in efforts to regenerate the natural defenses of their small island homes. At the time of my research, they were in the early phases of replanting coral in Villingili, inspired by others who are working to save reef systems throughout the world. Thanzy explained that "a lot of divers and environmentalists want to help out, who have been doing work since before we were born, people who were ready to give up, who are now back on track."

For Thanzy, placekeeping entails efforts to keep the island ecology going and is being negotiated between divergent environmental values and political futures. Through the work of Save the Beach Maldives, which has taken shape in the shadow of the seawall, place and belonging are being redefined to account for the natural environment and the ecologically diverse islands of the country. The efforts of Save the Beach Maldives to regenerate coral are deeply rooted in the dream of returning island ecosystems to their former glory. As they experiment with coral planting, others are expanding into new and far-reaching terrains, including the work of Reef Design Lab in Australia to build artificial reefs. Artificial reefs are often built from concrete, modeled on natural reef assemblages through a process of 3D printing.

The drive to create artificial reefs stems from the recognition that coral is in global decline and that this poses dangers for both the natural world and humans. However, the construction of artificial reefs is contextualized by place-specific concerns and politics. In the Maldives, the hope for artificial reefs is caught in the dilemma of preservation and the need for protection against sea change, including the damage wrought by decades of urban development. For this reason, coral planting is not only pursued for environmentalist purposes; artificial reefs also hold the promise of regenerating natural protections for inhabited coastal environments—to fashion marine-based *living* seawalls in place of terrestrial concrete structures that embody authoritarian configurations of power. The term "living seawalls" is relatively new and is associated with the process of affixing textured ceramic tiles onto existing seawalls to attract flora and fauna and bring life back to distressed ocean ecosystems. A broader conception of living seawalls includes the regeneration of natural barriers and allows for the multiplicity of ways that people reimagine and take ownership of coastal protection to make for a better world.[35]

The work of Save the Beach Maldives is one of the many outcomes of the democratic uprising that emerged in the aftermath of the 2004 tsunami and echoes the MDP's plans for resilient island planning and soft adaptation. However, as Thanzy cautions, development is part of a process that feeds on environmental disruption:

> I appreciate what Anni [Nasheed] has done, and I think it's important that there be a voice for small islands like us. It's quite sad that we have to be the people who are going to be first impacted by this when we contributed least to this problem. It's not a fair world. It's also a vicious cycle, because people in the poorer countries are forever working to become rich, and how do you become rich? By not being considerate with the world's resources and exploiting them.

While it is widely assumed that the decline of coral is primarily related to climate change and ocean acidification, local anthropogenic pressures have also played a significant role. Many of the natural sea defenses in the Maldives have been undermined by development, contributing to coastal vulnerability as well as food scarcity, a combined threat to traditional livelihoods in the Maldives. Villingili's outlying coral reef is severely damaged—and according to Save the Beach Maldives, it is "completely dead." Without a functioning and healthy coral ecosystem, the capacity of many of the smaller Maldivian islands to adapt naturally to rising sea levels is greatly diminished. Living coral, which is connected to the production of sand, has the capacity to adapt to changes in sea level. When corals die, the supply of sand decreases, and as they crumble, they can no longer protect islands from storm surges, leading to erosion.[36]

Bluepeace, an environmental NGO in the Maldives that protects lagoons from development, adds that "coastal development such as land reclamation and increased harbor dredging is the major and most common human-induced physical danger to coastal and coral reef ecosystems in the Maldives."[37] Bluepeace criticizes the Gayoom government for pushing to create "an artificial paradise on Earth," noting that such plans include channel blasting with dynamite, dredging, and reclamation—practices that have already undermined the health of the reefs, resulting in severe beach erosion.

Historically, Maldivians lived sustainably with coral, harnessing its gifts to craft monuments and houses of worship. Prior to modernization, Maldivians created small harbors for their fishing vessels by carefully repositioning coral in the reef. These practices involved a set of skills that allowed islanders to manage evolving island ecologies without outside influence. These initial breakwaters emerged alongside the creation of the *dhoni*, an Indigenous sea vessel designed to navigate the shallow waters of the atolls. To maneuver the coral, Maldivians assembled rafts of buoyant trees, attaching them to submerged corals before pulling sections into shallower areas.[38]

With modern development and the impulse for economic growth, coral was converted into a cheap and seemingly infinite building material for construction. In the 1970s, during one of the most rapid periods of modern development, coral was sourced from the reefs for land reclamation, often mined by blasting reefs with explosives.[39] The use of coral subsequently expanded into other projects, replacing traditional coconut wood, which had become scarce with the added pressure placed on the dwindling water supply. Coral was considered an ideal building material and became symbolic of conspicuous consumption in the Maldives. As Alexander Naylor writes:

> Coral construction allowed for better fire resistance, thicker walls and multistory buildings, key advantages in the increasingly-dense island settlements, especially Malé. When residential coral construction started to expand in

the early 1970s it became a sign of conspicuous consumption, with the historic use of coral in palace and mosque construction imparting coral with strong cultural connotations of stability. During the Malé Land Reclamation Project the annual amount of coral mined near Malé for construction purposes quadrupled from 1000 to 4000 m^3 between 1981 and 1985.[40]

The unsustainable and unregulated use of coral for development weakened many of the natural barriers that have protected inhabited islands for thousands of years, exposing newly developed areas to unprecedented dangers. In an interview, a government official explained how coastal vulnerability has increased over the past few decades as people have moved closer to the reef edge: "People occupied every piece of land available, and in some cases, they have reached to the greatest extent they can, and beyond that in the form of reclamation. That means, you have moved closer to the sea, of course the sea is also moving close to you, it's a bidirectional movement, and so you get more conflicts as a result." As people move closer to the edge of the reef, they settle near newly installed coastal infrastructure. A government official with the Ministry of Housing explained how this influences vulnerability: "I don't blame just the tsunami or sea level rise, there are other effects as well, development, harbors, access, jetties that also worsen our problems."

When the 1987 flood destroyed much of Malé, the damage was exacerbated by the loss of coral during rapid development. Naylor emphasizes the role that land filling and coral mining played in increasing the island's vulnerability to the flood: "The mining activities off of Malé reduced the height of the remaining reef by a half-meter, and the extension of land closer to the edge of Malé meant the new land was also closer to deeper water."[41] The cost of the damage ultimately exceeded the cost of the project. And while the 1987 flood prompted a ban on coral mining, builders illicitly gathered coral from the sea with little resistance from the government.[42] It was only after the government subsidized concrete that the use of coral slowed.

Concrete seawalls have also contributed to the decline of coral. Much of this stress has come in the wake of Gayoom's development program, which prioritized the construction of seawalls, harbors, and other hard structures that have accompanied the dredging of lagoons and the clearing of vegetation from beaches. Once island dynamics are locked in place by concrete, they cannot respond to natural changes, and thus they lose their adaptive capacity. Paul Kench explained this in an interview: "What happens is that the sand can't get around, so it all builds up. And then the next monsoon, this sand disappears pretty quickly, and then you start to erode the land." Thus the integrity and longevity of developed islands rests on the continued construction and maintenance of man-made structures, until the money, or the political will, runs dry. The resulting disruption has given rise to unprecedented coastal vulnerability.

Living seawalls, as an alternative to terrestrial concrete seawalls, join a new trend in architectural design that marks a shift from conventional land-based infrastructure toward what is now being called aquatic infrastructure, or "aquatecture." With the onset of climate-related flooding and attention to the vulnerability of land-based infrastructure, aquatic infrastructure has garnered a lot of interest from small island countries and international funding agencies alike. However, Asturo Morita, who studies the social implications of aquatic and terrestrial infrastructures in the Chao Phraya Delta in Thailand, argues that water-based infrastructure is not new but rather fundamental to traditional lifeways that were challenged and often erased by modernization.[43] Morita sees aquatic infrastructure as representative of "an indigenous form of urban space that is adaptive to dynamic changes in water flow."[44] He defines aquatic infrastructure as traditional canal transport and flood-resilient housing; terrestrial infrastructure emerged later as a by-product of modernization to enable irrigation and car transport.[45]

In the Maldives, the construction of living seawalls is an informal practice that is still in its early phases of testing. The appeal is easy to understand. Flying over the smaller islands of the Maldives, you can see the damage of land reclamation. In 2019, I spoke with Alex Goad, founder of the Australian-based Reef Design Lab, about how artificial reefs and living seawalls are envisioned in the Maldives. Alex was quick to qualify that these are conceptual projects, and wanted to ensure that their possibilities are not overexaggerated.

> Obviously, a 3D printed reef is never going to be able to replace something as grand as the Great Barrier Reef. So, within what we're doing, it needs to be clear that these aren't solutions in any sense for a lot of the problems that we're facing, especially with coral reefs. If we don't move to renewable

> energy and stop putting CO_2 in the air, we are going to lose the environment. It doesn't matter how many 3D printed reefs you put in place or even how many coral farming projects are put in place, these ecosystems just cannot be rebuilt.

The hope of a living seawall is a practical dream born from decades of disruption and the sobering realization that in order to bring life back to damaged island ecosystems, some form of technology is probably necessary. Reef Design Lab works at the interface of design and urban renewal, merging old ideas with new materials.[46] Its progenitors are young environmentalists and innovators seeking to stumble upon scalable solutions. They are excited about the collaboration that takes place in the design and construction of the molds and materials used for regeneration. These collaborations, between scientists, designers, engineers, public policy officials, and industry leaders, offer more hope than the devices themselves.

ATTACHMENTS TO CONCRETE AND THE LURE OF NATURE

Living seawalls and other green designs do not always challenge the underlying social and environmental injustices that create conditions of vulnerability and inequity. While some see artificial reefs as a means of negotiating with and regenerating their dying homelands, others see them as a lucrative way to maintain a steady flow of tourists to an increasingly artificial paradise fashioned in the context of declining natural reefs and fish populations.[47] Developers see potential for maintaining the aesthetics of pristine island landscapes through the use of these hybrid structures that offer some of the benefits of hard, resistant structures while maintaining the appearance of softer and more naturalistic ones. Artificial reefs fit nicely into the illusion of what William Cronon terms the "myth of pristine nature,"[48] and what Uma Kothari and Alex Arnall term "touristic nature."[49] Such expectations are rooted in the persistent imaginary of colonial aesthetics, translating the illusion of untouched paradise into a commodity for the taking.[50]

With the emerging popularity of green infrastructure, Gayoom's population consolidation plan is easily rebranded to embrace the environmental ideals of green capitalism. In a recent paper from the fields of architecture and construction, Ahdha Moosa and colleagues from Curtin University in Australia offer aquatic architecture in the form of floating cities as an adaptation pathway for the Maldives, curiously inspired by the discourse of "urban resilience" and the US-based libertarian Seasteading movement.[51] Moosa argues that such a pathway would "provide an adaptive sanctuary migration and habitation for populations threatened by RSL [rising sea levels]" while posing the question: "Should we fortify and hide behind ever-growing walls, retreat from the rising tides, or adapt and shape our cities into livable, responsive landscapes that engage and utilize the environmental context?"[52]

The desire for green infrastructure as a solution to climate vulnerability is leading to what is now being recognized as climate gentrification.[53] Soon after Moosa's paper was published, Dutch Docklands International, a global leader in floating infrastructure, revealed plans in cooperation with the Government of the Maldives to build a floating city. This new class of designer islands does not require land reclamation and is instead envisioned as the construction of interlocking floating cities that resemble coral in their designs, protected by the growth of new reefs. While floating cities might present themselves as the perfect middle ground between hard and soft measures, these, and all "solutions" to climate change, are bound to intersect with existing justice struggles on the ground. Furthermore, the extent to which these visions can take hold depends in large part on the whims of international funding agencies. This is evident in the lingering and unfinished traces of population consolidation and its competing vision of interisland transport and regeneration.

While green alternatives are gaining popularity, traditional hard structures are not easily abandoned. Regardless of the promise of green infrastructure and living seawalls, the preference for concrete seawalls remains strong in the Maldives, even as they contribute to spatial inequality and uneven development. Despite the fact that Gayoom's Safer Islands Development Program is despised by many Maldivians, Nasheed's soft adaptation alternative was not easy for small islanders to accept, especially those confronted with the immediate threat of erosion. The seawall that enabled the expansion in Malé has contributed to what Patrick Nunn, who studies Pacific islanders, calls a "seawall mindset," defined as a strong preference for hard structures among islanders.[54] Given that seawalls line the wealthiest parts of the world, it is not surprising that those who suffer from erosion—no matter the causes—would desire the privileged technologies of the developed world, including its seawalls.

Jon Shenk's film, *The Island President*, illustrates some of the complicated feelings that islanders who live outside of Malé have about seawalls.[55] In one scene, Nasheed is escorted by villagers to the edge of an island where erosion is taking place. As he makes his way to the shore, waves lap against broken stones. Nasheed pauses to stand atop a riprap structure, gazing into the ocean, perhaps contemplating a solution to the crisis unfolding under his feet. Later, Shenk's film crew follows Shauna Aminath as she travels from Malé to her home island of Hithadhoo in Addu Atoll. The scene shows Shauna with her family at a beach that is eroding. "This wasn't like this before," they tell her. "Part of my job is trying to prevent us from sinking," Shauna explains. They laugh and reply, "You mean, save us from sinking into the ocean?" Another asks, "So why don't you build a seawall? We should build a seawall around the entire island." As Shauna tells the women that the answer is not to build more seawalls, the laughter fades. One replies, "It seems like the waves are going to wash us away."

Seawalls remain intimately tied to the dominant development paradigm in the Maldives and all over the world. Yet not all Maldivians view seawalls as necessary. Mohamed Aslam stresses that seawalls are futile in the face of climate change. He likens seawalls to fences: "A fence is not going to save you. Just putting a wall around your city is not going to protect you." A Maldivian environmentalist who arranges field trips for children to spend time outside of Malé advocates for leaving islands untouched:

> It's a loss to our natural beauty. . . . Here in Malé you cannot see a turtle that can come to lay eggs because it's surrounded by concrete walls. . . . And if we do this to each and every island? I believe there must be islands free from any structure, any development. Total wilderness. *Leave it alone.* It is not necessary to take all these thousand islands for our uses.

The argument against traditional concrete seawalls is not only ecological; it is also social. For those who are opposed to population consolidation as an authoritarian form of managed retreat, alternative forms of protection and resilience would enable a future beyond centralizing infrastructures as the only means of staying in place, introducing smaller-scale protections and enhanced mobility for a diversity of islands. Paul Kench, who stresses the geologic importance of leaving islands free from development, finds the preference for hard structures discouraging and blames international funding agencies: "What disappoints me is that adaptation is still synonymous with physical structures and it's very hard to see that changing because a lot of the donor-driven money wants to see actual works done." According to Kench, there are alternatives to building more seawalls, but convincing the public would require "actually showing people successful examples of other ways of doing things."

For many Maldivians, the urgency of erosion and displacement cannot wait for politicians. Thanzy points to the historical responsibility of industrialized countries: "I've been told that a problem before the sea level rise would be ocean acidification. Either way, we lose our coral reefs, we lose our fisheries, we lose our way of life. So I suppose there is no justice really." At the same time, she feels inspired by movements for social change: "But there's been revolutions in history. So it can happen. And it's always been the people. So surely something can be done."

CONCLUSION

While perhaps a small drop in the ocean, the Maldives is a place of substantial dilemmas. The story of the Maldives invites us to think about what counts as adaptation and who gets to define its purpose—whether it involves seawalls, artificial coral, or floating cities. As one adaptation scheme embraces seawalls through population consolidation, pressuring people to move away from their ancestral homelands to fortified "safe" islands, another aims to regenerate islands and expand interisland transport to connect islands together. Across both of these visions, sensationalized narratives of climate change operate, sometimes legitimating lesser-known environmental injustices.

Staying in place is something that Maldivians strongly desire, and as such, placekeeping involves negotiating with who the country is for. This dilemma is further complicated by the economic logics of development and competing efforts to define history and what constitutes Maldivian culture, and therefore to define what is worth saving.[56] Coupled with the political struggle to define the future of the Maldives are efforts to preserve ways of life that are themselves contested. As small island cultures disappear to forces beyond climate change, what will be left to save? While internal migration has long been a necessity for Maldivians in the ever-changing island ecosystems of the Maldives, the dramatic shift from traditional small island life to an urban existence based on modern designer islands is an entirely new expectation—one that is directly imposed by Gayoom's population consolidation plan. And in the urgency to protect the Maldives from climate change, international funding agencies—guided by the principles of development—have lost sight of what is actually being protected.

The dream of "Aneh Dhivehi Raajje" and efforts to save the beaches and regenerate coral are but two examples of how Maldivian people are fighting back against the tides of disruption and injustice to stay in place. This desire, present in so many efforts to bolster barriers, natural or artificial, is a major driver in the contested politics of adaptation, not only in the Maldives, but throughout the world. While living seawalls are hardly an answer to climate disruption, they do tell a cautious story of empowerment and coexistence—where artificial reefs carry the hope and desire of bringing life back to the Maldives. Yet without a firm commitment to justice, adaptation in the Maldives is bound to mirror histories of unjust and uneven development.

Conclusion

THE DILEMMA OF PLACEKEEPING

At the 2014 United Nations Climate Summit in New York City, in a building slated for protection by a massive seawall, Kathy Jetñil-Kijiner, a poet and activist from the Marshall Islands, gave expression to the existential peril of climate disruption. Her words, composed as a letter to her newborn daughter, echoed far and wide as she spoke of the "lucid, sleepy, lagoon lounging against the sunrise" that might one day transform into a monster that would "gnaw at the shoreline, chew at the roots of your breadfruit trees, gulp down rows of your seawalls and crunch your island's shattered bones."[1]

As she urged her baby not to cry, the dark image of coastal injustice gave rise to a conviction to stay in place—to fight back against a future where the people of low-lying and small island nations are forced to "wander rootless with only a passport to call home." "Don't cry," Kathy said, smiling. "Mommy promises you no one will come and devour you. No one's drowning, baby, no one's moving, no one's losing their homeland. . . . We are drawing the line here."

The desire to stay in place is deeply felt across the many front lines of climate change, prompting ordinary people to take action in extraordinary ways. In the Maldives and Guyana, where destructive development has locked a generation behind a seawall, drawing the line is an act that engages with intersecting struggles for justice. This means contending with racial injustice, histories of colonization, and practices of authoritarian development. It also means negotiating with mobility justice and the right of home, freedom, and access to public goods and resources rather than walls that restrict people to new configurations of capitalist development. These struggles are at the root of coastal and climate justice: the right to choose, to be free of economic dependency and the trap of debt and the burden of being faced with false choices and double standards.

Existing frameworks of climate adaptation fail to capture the intersecting injustices and inequalities that exist within and between places. Coastal disruption is a process enacted over time and through multiple and overlapping waves of injustice. Efforts to open access to the ocean and expand the reach of human dominance have created the conditions of vulnerability experienced in Guyana and the Maldives. Adaptation is a predominantly top-down, expert-driven concept that overlooks place-specific histories of dispossession, political oppression, and everyday efforts to stay in place. It also obscures global injustices and histories of colonization and development in favor of solutions that serve the interests of investors and privileged members of society. Adaptation is often framed to accommodate future threats at the scale of individual communities, failing to consider the larger systemic forces that have contributed to coastal disruption and conditions of vulnerability over time. Without attention to the social complexities of climate vulnerability and the hidden struggles of *placekeeping*, adaptation is bound to exacerbate existing global inequalities.

As environmental values change in response to the destruction of critical ecologies, adaptation is at a crossroads. Competing logics of coastal protection are now being negotiated alongside struggles to define and maintain a sense of place and belonging in areas that are exposed to multiple and overlapping threats. Sustainable pathways exist in complicated relationships to the physical and social assemblages of gray infrastructure installed over decades and sometimes centuries. While green pathways are possibly more equitable, they are not perfect solutions.

The stories of Guyana and the Maldives provide valuable insight into the specific social dilemmas and consequences suffered in relation to the obstruction of shoreline ecosystems as well as the unanticipated avenues that people must pursue in response to the construction of seawalls. Below

I highlight some key points that I hope will inspire new and more critical ways to approach the problem of sea change from a place-sensitive perspective.

IN THE SHADOW OF THE SEAWALL

In the shadow of the seawall, we see that placekeeping can entail many actions, from women safeguarding mangrove forests to democratic uprisings and youth movements to save the beaches. We also see that seawalls embody politics and histories of oppression. Zooming out, it becomes clear that seawalls play a role in shaping the contours of sea change and global inequality. Yet seawalls are also boundaries around which livelihoods and struggles to stay in place are being negotiated. They are becoming sites in which people are fighting for their rights and for their vision of a better world. Together, all of these insights point to the importance of placekeeping as a critical intervention in the study of climate adaptation.

Guyana and the Maldives illustrate the profound importance of place in the battle to stay above water. Together, they fill in gaps that are missing from current adaptation frameworks and open new questions for how we are to understand the future, demonstrating the salience of placekeeping as a process that involves maintaining livelihoods while negotiating with oppression, attachments to place, and the anxiety of anticipated loss. These two seawall entanglements—one sunken within a dense alluvial wilderness and the other floating atop a glimmering coral reef—embody relationships of power that mirror long-standing struggles for sovereignty and democracy in which competing logics of adaptation are being tested and negotiated alongside the assumptions and political motivations that accompany them. Together, they blur the line between dichotomies of local/global, hard/soft, and good/bad while showing us that wherever a seawall exists, the dilemma of placekeeping is present.

Seawalls as Entanglements of Political Desires

Seawall entanglements in Guyana and the Maldives exemplify the dynamic role of infrastructure in the making and remaking of place and also in the production and remediation of coastal vulnerability. Together, the stories of Guyana and the Maldives remind us of how the past becomes entangled in the present through infrastructural assemblages that are more than the sum of their parts; they engage with place-based politics that complicate adaptation frameworks. In these cases, it becomes clear that seawalls enter into struggles for democracy and sovereignty in contexts of unjust land use practices and political oppression. They embody relationships of power that work to extract, commodify, and centralize development. While much of this transformation was deliberate, seawalls also possess what Langdon Winner terms *inherent politics*.[2] This means that the very design and construction of seawalls accompanies hierarchies of knowledge and planning that align with oppressive and authoritarian relationships of power that continue even after those in power have faded.

Waves of colonization, development, and globalization have transformed shorelines into vulnerable places, creating cycles of creation and destruction. These waves have destroyed the planet while also offering remedies through measures that aim to protect the environment and maintain shorelines as they existed prior to disruption. This drive to maintain the baselines of modernity underlies a paradox of permanence that continues through sustainable development and the more recent turn to green infrastructure. Seawalls, for example, create the conditions of their own vulnerability and carry forward politics that seek to transform coastal environments into places of social control, hegemony, and power.

In Guyana, seawall formations originated in colonial conquest. Accumulated over thousands of years in the Frisian lowlands, Dutch knowledge was used to tame the Guyanese coastline and to yield sugarcane for colonial extraction. As a result, the Indigenous Amerindian population lost access to the coast. Enslaved persons and indentured servants were violently transplanted into the unforgiving coastal marshland. When Guyana changed hands to British rule, seawalls were constructed to further expand colonial plantation economies and their accompanying infrastructures of flood control. Seawalls were used to divide the land for extraction based on the knowledge of early colonial coastal engineers, who introduced concrete to the Guyanese coastline.

Fortifying plantations was an expensive endeavor, and seawalls were central to the formation of racialized class division that locked free African and Indian wage earners, Indian indentured servants, and peasant farmers into cycles of poverty. Walter Rodney referred to this as the emergence of the planter class, a divide that was exacerbated under the rule of Forbes Burnham after Guyana was liberated. Funding to protect Guyana's shore further politicized the role of seawalls. After colonization, funding came from the United States, a rising superpower that aimed to eradicate communism from South America. Later, funding came from international organizations guided by development logics and cost-benefit objectives that were incompatible with Guyana's unique ecological and social landscape.

In the Maldives, this relationship is evident in the centralization of development and the role that seawalls have played in reshaping and restructuring islands for urbanization while displacing Maldivians from their ancestral islands. While the history of the Maldives is quite different from that of Guyana, both countries have experienced colonial disruption. The unique landscape and distant location of the Maldives largely protected the country during the height of European colonization but not entirely. The Maldives has a long history of monarchy and centralized power that aligned with colonial powers. The political elite of the Maldives gathered momentum in the era of development. International development agencies enabled powerful, autocratic rulers to reshape landscapes through modern infrastructure projects.

Seawalls in the Maldives are accompanied by a politics of authoritarian rule and uneven development, a reality that is easily visible in the extremes of spatial inequality. Since the first modern seawall was constructed by Japan in the 1980s, the Maldives has become deeply divided by land reclamation projects and artificial islands that centralize resources and rely on the construction of additional seawalls. These projects have challenged an attachment to place that involves traditional livelihoods. Small islanders are already at risk of being lost before climate change enters the picture.

In both Guyana and the Maldives, seawalls have also contributed to an unsustainable maintenance burden that fuels relationships of dependency and international debt. As large-scale infrastructure systems are taken up by centralized governing bodies, resources are siphoned to the development of projects that must be maintained indefinitely. This creates overwhelming challenges for new and more democratic political futures. As was seen in the case of the Maldives, the infrastructural entanglements of the past come into conflict with efforts to envision a different future. The high cost of seawalls solidified an authoritarian vision of centralized power that continued to place demands on the democratic government that followed in its place. With the construction and maintenance of seawalls requiring nearly a third of the national income, the burden of maintaining seawalls limited the possibilities of alternative development priorities. Mohamed Nasheed's vision for affordable interisland transit and soft adaptation was a courageous attempt to shift the Maldives in a new direction but was quickly silenced by political repression. Like a pendulum, adaptation shifts between gray and green infrastructure as democratic and authoritarian leaders vie for power.

This suggests that seawall formations call into play authoritarian relationships of power and hierarchies of knowledge and funding. The political nature of seawalls in Guyana and the Maldives is not contingent on their builders; the politics live on through the demands of maintenance. Over their lifetime, seawalls embody expert knowledge, top-down planning, and the political agendas of powerful international agencies. When seawalls come into crisis, experts are called in to manage them, reflecting the lingering influence of colonial power and knowledge. In both Guyana and the Maldives, conflicts between "expert" and local knowledge speak to the lasting impacts of the colonial mind-set that justified the transformation of the coast. However, just as seawalls are never fully permanent, their politics are not fully concretized. While seawalls have inherent politics, challenging these politics is possible, and people are doing it.

Seawalls as Sites of Negotiation

The widespread construction of seawalls and other hard structures draws attention to an uneasy contradiction between desires for permanence, environmental sustainability, and struggles for a better life. But rather than abandon seawalls, there is a new drive to make seawalls "green" through techniques of biomimicry and the operationalization of natural metaphors for coastal protection. Seawall entanglements in the Maldives and Guyana accompany contested politics and divergent visions of the future. Coral planting, which often utilizes molds and artificial reef designs and principles, is not unlike the work of mangrove regeneration. The search for alternatives and hybrid approaches invites tension, but placekeeping is more than staying above water; it is a struggle to maintain a culture and define what is worth saving.

The environmental turn has inspired new ideas for staying in place, but these new pathways must negotiate with the physical and cultural assemblages of the past. They exist, as Susan Leigh Star notes, on an installed base.[3] In the context of coastal disruption and the anticipation of rising sea levels, these values are being negotiated in line with the inherent politics of seawalls. As such, seawalls have become sites of negotiation where diverse sentiments, memories, and livelihoods stand at the center of competing desires for permanence.

Acts of placekeeping in Guyana and the Maldives make it clear that people are not passive recipients of oppression. Nor do they sit by idly as their environments are transformed into unsustainable and unlivable places by the forces of colonization, development, and now climate change. In Guyana and the Maldives, this has translated into political movements for democracy that coalesce with efforts to regenerate the natural barriers that stabilized the coastline long before concrete was introduced. These efforts are challenging the inherent politics of existing concrete seawalls while embracing innovative and informal approaches to the dilemma of staying in place. Decentralization and the struggle for democratization and small island autonomy in the Maldives mirrors the efforts of mangrove planters in Guyana, where placekeeping entails confronting the impossible burden of maintenance to find independence and sustainability. Across both cases, as people work to maintain a sense of place, they are actually working against waves of political oppression.

Seawalls are large-scale projects that rely on top-down planning. Placekeeping, on the other hand, occurs on much smaller scales specific to the unique needs of particular communities. As development, erosion, and rising sea levels threaten coastal areas in Guyana and entire islands in the Maldives, placekeeping often starts with unmet needs and a perceived gap in governmental support. In the case of Guyana, the organizers of the Guyana Mangrove Restoration Project felt that the government was not doing enough to protect the coastline or to provide for marginalized members of society, single mothers in particular. In the Maldives, community organizers felt that government projects were leading directly to the decline of coral. Placekeeping started in both places as an informal effort that then found support through organizing and international funding.

With the shift to a desire for green infrastructure, new seawall formations are emerging that embody politics of resistance and regeneration. While this has translated into more democratic values in both Guyana and the Maldives through the use of mangroves and coral regeneration, these politics are becoming increasingly embraced by powerful players from within the same development models that brought waves of disruption to the shore. In Guyana, planting green seawalls is both a grassroots effort and an effort endorsed by international funding agencies and a new wave of adaptation planning. In the Maldives, soft adaptation plans were managed in a top-down manner, although combined with a grassroots effort to decentralize infrastructure.

Negotiating with the infrastructural entanglements of the past is central to the story of coastal Guyana, where the regeneration of mangrove forests involves envisioning a future that diverges from the lingering ruins of colonization. In this context, regeneration carries multiple meanings. Regeneration is about preserving existing forests in decline for the purposes of adaptation, but perhaps more importantly, regeneration is an act of placekeeping intended to resurrect and create new landforms in places that have been made vulnerable by colonial transformation and climate disruption. The promise of regeneration is the hope of autonomy and freedom from centralized infrastructural systems of authoritarianism and dependency. However, this vision is not shared by everyone and challenges a long-held sense of place that is intimately tied to ocean breezes and a visible shoreline.

As new possibilities emerge, desires for concrete seawalls are not easily abandoned, even as projects of coastal fortification squeeze the shore, limit other possibilities, and threaten to displace people. In Guyana, when engineers made the decision to replace many of the declining colonial seawalls with riprap structures, the shift to riprap was perceived as an example of government corruption. A lifetime of colonial seawalls, combined with memories of better times, complicated public perceptions of seawall alternatives. Similarly, in the Maldives, many small islanders facing the immediate danger of erosion hope for a seawall. This is not because they hold special memories of seawalls; rather, they see that other islands with the privilege of seawalls are faring better in the landscape of uneven development.

Like texts, the value and meaning of seawalls can change over time. When coupled with discourses of adaptation, seawalls can obscure darker motivations. In the Global South, unjust development priorities are increasingly being reframed as national climate adaptation plans in order to secure funding and support from international lending agencies. For example, in the Maldives, the discourse of climate change adaptation is being used to further justify the proliferation of mega seawalls and artificial islands that are being used to phase out small island life. Unpopular relocation schemes are repackaged as central to national resilience plans. Often, these messages have little to do with environmental values and more to do with the pursuit of economic dominance in a global economy that thrives on inequality. For instance, the search for oil in both Guyana and the Maldives reveals how easily resilience discourses are abandoned by those in power when fossil fuel wealth is within reach. And when these narratives are exposed as exaggerations that deny the full context of vulnerability, they can serve to legitimate the claims of climate denialists.

SEAWALLS AND CLIMATE APARTHEID

In *The Rising Sea*, Orin Pilkey and Rob Young describe the relocation of the tiny village of Shishmaref, Alaska, as a calculus of multiple options, each involving unique social and economic costs.[4] The first alternative is to build seawalls, which is the least expensive option, but for Shishmaref, building seawalls in the shifting permafrost introduces many untenable problems. The remaining alternatives involve some form of relocation but at different distances and with associated social and cultural repercussions. They involve moving slightly more inland, moving to nearby villages, moving to the nearest large city, or moving the community as a whole to the mainland. The favored option is to move Shishmaref to the mainland. This is also the most expensive option, and Pilkey and Young acknowledge that such a project is not viable for all frontline communities. Unlike communities in the Global South, which are "virtually on their own," Shishmaref's survival is "backed by the deep pockets of a rich country."[5]

As measures of last resort in the struggle to hold the line, seawalls have become symbols of uneven protection. With adaptation measures favoring the interests of the powerful and wealthy, the world is headed in the direction of climate apartheid.[6] In the existing context of rampant global inequality and uneven development, what might seawalls tell us about the future? Will we face a world in which windows are needed to look out to the ocean, as is already the case in parts of Japan? Will the absence of a seawall commemorate a victory for environmental conservation and climate justice—or will such an absence signify another site of injustice on an increasingly damaged planet?

The contours of Earth's coastlines are being reimagined in the context of sea change, and so too are the sacrifice zones. With increasing urgency, seawalls are merging into changing coastal landscapes, mirroring relationships of global inequality as they work to protect areas against sea level rise.[7] The struggle to find permanence amid conflicting financial and ecological calculations of local, national, and international development priorities and concerns over sea change cannot be separated from contexts of global inequality. Global inequality, the result of multiple and overlapping waves of injustice, started with the rise of global capitalism, the expansion of colonial rule, nation-building projects, industrial production, development, resource depletion, and the intensification of war. Neocolonial adaptation schemes now threaten to further endanger coastal communities as coastal protection and managed retreat mutually contribute to forced removal.

The cases of Guyana and the Maldives clarify the connection between uneven development and uneven adaptation, and this pattern is bound to be repeated on a global scale as the wealthy are afforded their own protections. In the Maldives, uneven development became synonymous with adaptation through an authoritarian plan to consolidate the Maldivian population to a handful of "safe" islands encircled by seawalls. In Guyana, colonial extraction led to uneven development, locking 90 percent of a divided population behind a patchwork of decaying seawalls. Both of these seawall entanglements were supported by developed countries and international funders without a commitment to maintain coastal protection through the inevitable trajectory of erosion, decline, and climate change. Meanwhile, wealthy countries are enacting their own plans for protection while encouraging poor countries to find alternative, green solutions.

While spending time in Guyana and the Maldives, I also encountered seawalls forming in wealthier parts of the world, where similar logics are operating but with different standards. In the context of the climate crisis, seawalls are already being used to construct enclaves for the wealthy. Two years after Hurricane Sandy, a contract of $335 million was awarded to an interdisciplinary team of scientists, engineers, designers, and architects to build the Big U, a ten-mile seawall around the Manhattan Financial District. Made possible with disaster relief funds, the plan is one of the largest coastal protection projects in New York history. Under the banner "Rebuild by Design," the architects proposed a structure that would blend into the built environment, an invisible shield apparent only during floods.[8] The project description mixes narratives of environmental sustainability, where people are able to "farm, sunbathe, eat, and engage with world-class gardens," with those of community-oriented design aimed at "enhancing the public realm while protecting the Financial District and critical transportation infrastructure."[9]

Even among the wealthiest places in the world, the promise of protection is never certain. As plans for protecting New York with a seawall materialize, engaging the sublime vision of the Netherlands to guide much of the dreamscape of coastal protection, they invite many unforeseen challenges and contradictions. The plan, initially engaged through a participatory community process, was later overturned by a top-down design endorsed by city officials. The new plan called for traditional hard seawalls rather than a solution that would provide for multipurpose space, moving the project from its earlier community design principles and opening the floodgates for green gentrification.[10]

I witnessed the construction of a similar enclave of coastal protection in Venice, where I took a detour between my travels to Guyana and the Maldives. Shortly after Hurricane Sandy, the *New York Times* published an article comparing the situation in New York to the "mammoth" seawalls of Venice, questioning whether the hard path was the most appropriate option.[11] The barrier walls of Venice (more widely known as floodgates) are a perfect example of how futile and politically divided such projects can be.

Built to protect the coast and to expand and reconfigure the landscape for tourism, the MOSE Project, named after the biblical Moses and the acronym MOdulo Sperimentale Elettromeccanico (experimental electromechanical module), is the culmination of over thirty years of negotiations set against a backdrop of economic crisis and political scandal. At a cost of $9.5 billion, MOSE is literally designed to part the waters, consisting of a system of mobile gates positioned at each of three inlets separating the Adriatic Sea from the Venetian Lagoon. The gates extend more than a mile, each component measuring up to 92 feet long and 65 feet wide and weighing approximately 300 tons.[12] They are designed to lay flat below the surface of the water, out of sight. When triggered, the massive panels rise vertically on a hinge to form a wall that blocks the infamous *aqua alta* (high tide) from entering the Lagoon.

Despite the technologically advanced nature of the MOSE project, there are indications that it may not work as planned, suggesting that money doesn't guarantee survival in any enduring or holistic way. MOSE simply buys time. In Venice, it is widely claimed by the government that MOSE is climate-proof. This is a risky and perhaps irresponsible assumption, because while MOSE may hypothetically protect Venice from floods in the short term, it is not the sole solution to the ecological decline that threatens the continued existence of Venice.[13] Assuming that MOSE works effectively over many years, the threat of catastrophe still looms. When the construction of flood prevention structures encourages additional development, it often leads to an exponential increase in the damages suffered when the infrastructure fails. This is known as the "levee effect."[14] In "Things Fall Apart," Benjamin Sims reflects on the failure of the levees in New Orleans during Hurricane Katrina, which he argues has resulted in "deflated assumptions about the ability of a wealthy, technologically advanced society to handle a natural disaster of terrible, but not necessarily cataclysmic, proportions."[15]

For those Venetians who now find themselves backed against a wall over which they had little say, a different kind of injustice plays out.[16] In 2012, Anna Somers Cocks and six trustees dramatically resigned from the Venice in Peril Fund, founded in 1966 to restore monuments and works of art. The political corruption inherent to the development of MOSE had reached a boiling point.[17] Who suffers from this injustice? According to Cocks and Lidia Panzeri, "The Italian tax payer is the loser, but especially the city of Venice, which depends on money voted under the so-called Special Law for its exceptional and unavoidable costs, such as dredging the canals and repairing the ancient buildings."[18] Given the lack of political integrity and the alleged shortsightedness of the government, one begins to wonder whether these submerged seawalls will actually protect the city. With the Italian government already cast under a shadow of doubt, what faith is there for the success of a project like MOSE? What does this suggest about the Big U in New York?

Constructing seawalls around the most privileged places on Earth, including Manhattan and Venice, would not be possible without the bounty of wealth that was accumulated in contexts of global inequality. This reality is at the core of climate justice, where the ecological challenges faced by New York and Venice are in large part shared by the vulnerable poor who have been stripped of their resources and means to survive a changing climate for which they bear little or no responsibility. However, like most walls, seawalls do not always conform to the desires of their makers, and they can invite in what they are intended to keep out.

As the wealthy plot their path to preservation, the most vulnerable and disrupted areas of the world are hardly afforded such extravagant options. The seawalls of the future are wrought with double standards. For instance, the United Nations encourages poor countries to pursue soft, cost-effective measures to deal with the existential threat of sea change while its own structures are to be fortified with seawalls. The environmental sociologist Matthew Clement uses the phrase, "Let them build seawalls," to allude to the fact that the burden of adaptation is being transferred to those who have been exploited by the global economy.[19] When seawalls are financed in developing countries, they merge with existing injustices and contribute to spatial inequality and uneven protection, as was seen in the cases of Guyana and the Maldives. This is also the case in other parts of the world.[20]

This suggests that histories of uneven development are paving the way to futures of uneven adaptation. To challenge this cycle, movements for climate justice are not only demanding an end to fossil fuels; they are also starting to embrace a more critical and democratic process of adaptation planning. As a movement, climate justice calls for comprehensive revolutionary social change and embraces antiracist environmentalism, human rights, the rights of nature, and the principles of environmental justice that define the moral and ethical dimensions of social and environmental wrongdoing.[21] At its core, climate change is a crisis of human and societal relationships and a problem that requires comprehensive change at all levels of social life. It is also a crisis that stems from the dilemma of modernity and efforts to discipline and control the shore for the purposes of human settlement, industrialization, and development. Here the structural violence of poverty and the slow violence of climate change converge.

ADVANCING A PLACEKEEPING LENS

The making and *keeping* of place is both a cultural and a material process, involving ideas, values, and infrastructural interventions that come together to shape delicate physical and social ecosystems. A place-sensitive sociological perspective can help us see climate vulnerability as part of a larger process of placemaking in contexts of disrepair and disruption. Focusing on seawalls helped me untangle the many facets of climate disruption across waves of injustice and cycles of creation and destruction along the coast. Listening to the stories of people who live behind seawalls opened my eyes to the culturally specific ways that people engage in efforts to maintain their sense of home in politically and ecologically precarious landscapes. For them, place is both a physical location and a sense of belonging that undergirds the struggle to stay in place in the context of climate change. Places are not empty social canvases; they are embedded with meaning as well as desires for permanence that accompany infrastructural entanglements of power and inequality.

I believe this calls for new lines of inquiry that move beyond existing adaptation frameworks to address the nuanced dynamics of place and the struggle for permanence. I have offered *placekeeping* as an alternative framing for adaptation to emphasize the importance of people, ideas, and values in choosing and defining life-affirming pathways of existence. The injustices of colonization and modernization are not endured without resistance. For those who live with coastal disruption, placekeeping is a fight for home. Without attention to the physical objects, identities, livelihoods, politics, power dynamics, visions, vulnerabilities, and inequalities that constitute place, it is impossible to imagine and plan for fair and just adaptation.[22] Below I highlight some key insights for applying a place-keeping lens to the problem of climate disruption.

First, it is important to situate placekeeping as a cultural process that engages with multiple temporalities of vulnerability and injustice. The collective experiences of placekeeping in Guyana and the Maldives underline the need for a new understanding of coastal vulnerability based on an analysis of the diverse relationships that have come to exist across shorelines and technologies of coastal modification, from colonial encounters to schemes for economic development. This includes broadening adaptation framing to recognize that climate vulnerability engages with multiple injustices, taking into account the particular social and historical precedents that have contributed to the instability of shoreline landscapes and decisions to drain, dredge, and reclaim land. Understanding vulnerability in this way—as a place-specific process entangled in infrastructural baselines and global waves of injustice—allows us to see beyond the limited framing of conventional adaptation discourses and recognize when adaptation pathways conceal unjust development priorities that harm rather than help communities.

The stories described throughout these chapters also point to the importance of broadening notions of coastal justice beyond the single issue of beach access in order to consider how histories of injustice shape and give meaning to contemporary issues of coastal disruption. In fact, early struggles for coastal access and belonging emerged from racialized processes of exclusion and displacement.[23] Coastal justice began as a rallying call for marginalized and low-income communities displaced from wealthy, segregated shores.[24] In the context of climate change, coastal justice is now entwined with the larger concerns of climate disruption, shifting attention to rising sea levels and the temporal and spatial injustices of anthropogenic climate change. However, racialized histories remain embedded in the landscape, shaping conditions of vulnerability while complicating the process of adaptation planning. As Dean Hardy and colleagues caution, there is an urgent need to challenge the emergence of color-blind adaptation planning that ignores histories of racial injustice.[25]

This leads to a second critical dimension of placekeeping: staying in place is a relational struggle to define and create life-affirming pathways of survival. Staying in place is strongly preferred by those who are faced with the prospect of climate change, yet placekeeping is not a simple matter of staying or leaving. It is a relational process. The loss of place and the effects of displacement vary, and it is possible to become displaced without leaving.[26] Therefore, it is important to understand placekeeping in relation to the multifaceted threat of displacement and the various ways in which people can find themselves dislocated. This requires that we focus on the injustice of unequal mobility, as Mimi Sheller does in her work on postdisaster landscapes,[27] in addition to the displacement of rural livelihoods by adaptation planning that favors urbanization.[28]

Climate justice has the potential to embrace this intersectional and relational understanding of place. As Malini Ranganathan and Eve Bratman argue, climate justice activists are engaging with the rehumanizing goals of freedom and liberation through "place-specific ways of knowing and feeling."[29] They argue that in contrast to the semantics of resilience and adaptation, climate justice challenges the underlying forces "that impede the ability of people to lead healthful and dignified lives."[30] As it stands, climate justice on the shore focuses on issues of erosion, flooding, and loss of livelihoods, working to bring attention to the environmental injustices of runaway climate change.[31]

However, as notions of climate injustice give rise to the image of the climate change refugee, it is important to be wary of simplified narratives. Such depictions can serve to homogenize communities and deny agency to people who are not passive victims of climate change. While the refugee narrative is evocative and has been effective in providing a moral basis for the least developed countries to ask industrialized nations to change their behaviors, it obscures histories of internal displacement and processes of development that already operate to create conditions of vulnerability. Furthermore, misconceptions about how frontline communities negotiate with oppression turn into myths that limit our ability to visualize the nuances of adaptation. As Idowu Ajibade argues, this misreading of frontline communities "erases the innovation and cultural adaptiveness developed by poor communities over decades of living with water."[32]

The case of the Maldives suggests that there is a need to further broaden climate justice framings in order to consider how the right to stay in place comes into conflict with national adaptation measures that operate along existing lines of spatial inequality.[33] The struggle for small island justice in the Maldives means resisting adaptation plans that seek to concentrate populations and resources into a handful of urban islands with large seawalls. For many Maldivians, such plans run contrary to their desire to maintain their sense of belonging on small islands, threatening their livelihoods. This raises important questions about who adaptation is for and what forms of adaptation are equitable in contexts of uneven development and political oppression. Unjust development for the purposes of adaptation is a form of climate injustice.

The third point I wish to make is that a placekeeping lens requires us to challenge and remove unnecessary binaries that obscure everyday practices of adapting to change—for example, the "protect or retreat" dichotomy. While it is widely assumed that adaptation involves a set of definitive choices, the process is rarely experienced as a clean split between the options of coastal protection and managed retreat.[34] Rather, the two emerge and coalesce over time into a relational process. This combination of forces—coastal protection and managed retreat—cuts across both the Maldives and Guyana. At the level of planning, they might be perceived as mutually exclusive, but over time and across wider scales, they become mutually reinforcing practices that mirror existing structures of spatial injustice and global inequality. The ideological divide between protect and retreat perpetuates a false dichotomy that loses sight of the infinitely embedded social and spatial contexts of coastal disruption.

As managed retreat moves to the forefront of adaptation thinking, it is important to recognize that like coastal protection, managed retreat is guided by the same narrow cost-benefit logics that guide decisions about who and what gets moved, and to where.[35] In some cases, managed retreat is proposed as a strategy to avoid the higher costs of infrastructure protection. This can result in a host of social, economic, and psychological externalities of property acquisition, including loss of community, impacts to the local property-tax revenue, loss of sense of place, upending of the social composition of communities, loss of social support systems for those most vulnerable, and lack of livelihood opportunities in the place of relocation.[36] It can also lead to the unjust displacement of the poor, who have minimal power in decision-making processes and minimal resources for legal redress. These decisions often lack transparency and involve political and economic motivations based on cost-benefit analysis that prioritize coastal protection in wealthier communities while denying low-income or minority communities protection.[37]

The debate between managed retreat and coastal protection is both a financial contest and a negotiation over values about nature. Coastal protection is increasingly framed as a barrier to managed retreat, not only because of competing costs, but also because managed retreat is considered less destructive. However, coastal protection exists in a mutually reinforcing relationship with managed retreat, not in opposition. This was the case in the Maldives and can be seen throughout the developing world. For example, in Lagos, Nigeria, a large seawall referred to as "the Great Wall of Lagos" has reduced flooding in the city center while shifting erosion to nearby communities; meanwhile, the state government has enacted managed retreat schemes to remove informal settlers from desirable waterfront areas.[38]

A placekeeping lens invites us to ask critical questions about who is served by managed retreat or coastal protection, both before and after the process, and who has the power to make such decisions.[39] The underlying assumptions of any form of adaptation will favor particular groups, value systems, classes, or geographies. There is also a need to understand personal and situational contexts that drive decision making with regard to staying in place or relocating. While not relocating communities can be as harmful as forced relocation, there is a need for a socially just approach to adaptation centered on protecting livelihoods and empowering communities for the long term.[40]

THE SEAWALL AROUND US

In drawing this book to a close, I am left thinking about the words of the Old Man of the Wall from *The Seawall: Tales of the Guyana Coast*: "Sometimes it's like the sea is calling for sacrifices, as if the sea made this wall into an altar. What brings us here to this wall of dreams, of nightmares, mysteries, murders, and strange, strange happenings?"[41] This haunting sentiment captures the essence of what I see playing out in the shadow of the seawall. With a majority of the world's population living close to the shore, seawalls are not likely to vanish, even as environmental values shift to regeneration and green alternatives. The complexity of coastal vulnerability, rooted in histories of inequality, suggests that the proliferation of seawalls is our future—a future that is bound to be unstable, contentious, and unjust.

The stories of Guyana and the Maldives strongly suggest that seawalls are structures of colonial and authoritarian power, embodying politics that constrain future generations to hierarchies and dependencies that are not easily reconciled. In Guyana, seawalls tell of violence and brutality, forced labor, and stolen lives. In the Maldives, seawalls symbolize political oppression, rapid modernization, and forced relocation. Nonetheless, placekeeping alludes to a more hopeful observation that a different kind of politics is guiding the regeneration of natural barriers, hinting that another future is possible in which people are given the freedom to design and take ownership of their own protection measures.

Perhaps there are better—and more socially just—approaches to this problem such as transforming vulnerable areas into places of nonextractive living. As Naomi Klein argues, "This can go far beyond the usual calls for stronger seawalls: activists can demand everything from free, democratically controlled public transit, to more public housing along those transit lines, powered by community-controlled renewable energy—with the jobs created by this investment going to local workers and paying a living wage."[42] Andreas Malm wonders if perhaps seawalls themselves can embody nonextractive politics, asking, "Could there be a revolutionary sea wall?"[43] If so, he argues, such a seawall would aim to serve vulnerable people through regenerative and noncapitalist modes of creation. This raises an important question, posed by Melanie DuPuis and Miriam Greenberg, as to whether "the right to the resilient city" is achievable in contexts of economic inequality that are guided by logics of infinite growth.[44]

The nuances and competing visions of life behind seawalls, shaped by the frictions of race, class, gender, religion, environment, and culture, are not unique to the cases explored in this book. However, they are obscured by simplified narratives of climate change that avoid the complicated relationships inherent to our infrastructural assemblages. The historical and social inequalities that are woven into the built environment are often glossed over. Landscapes with existing seawalls are already physically and socially precarious, and in many cases, this points to former injustices that continue to shape struggles for permanence. Before the next seawall is built, it is worth asking what kind of politics our future seawalls will embody—symbolically and concretely—and how these politics will surround us in the present and eventually outlive us.

For frontline communities confronted with histories of environmental injustice and coastal disruption, staying in place is a desire that is burdened with difficult choices. This sentiment was beautifully captured in another poem by Kathy Jetñil-Kijiner, involving a collaboration with Aka Niviâna, a climate justice poet from Greenland. The poem, "Rise: From One Island to Another," describes the devastation of their geologically interconnected homelands, tracing the mutual destruction of melting ice caps to rising sea levels while emphasizing their shared conviction to fight back and stay in place.[45] In the closing verses, Kathy reflects on the invasion of climate adaptation that is "forcing land from an ancient, rising sea, forcing us to imagine turning ourselves to stone." "Despite everything," she says, "we will not leave. Instead, we will choose stone. We will choose to be rooted in this reef forever."

In Joanna Macy's book *Pass It On*, one story speaks indirectly to the ritual of permanence that characterizes the human condition. In telling of the rebuilding of a plundered Tibetan monastery at Khampagar, Macy writes:

> You do what you have to do. You put one stone on another and another on top of that. If the stones are knocked down, you begin again, because if you don't, nothing will get built. You persist. Through the vagaries of social events and the seesaw of government policies, you persist, because in the long run it is persistence that shapes the future.[46]

Similar to the interwoven stories of seawalls, mangroves, artificial islands, and coral reefs, these rituals of persistence are being negotiated alongside the changing political and ecological realities of the twenty-first century. Regardless of whether the future is one of more seawalls, artificial islands, or floating cities, the path always seems to be heading in the direction of more stone and more concrete.

Methodological Appendix

The empirical research highlighted throughout this book was gathered over four years between 2012 and 2016 during fieldwork in Guyana and the Maldives, as well as visits to the Netherlands and Japan to trace historical linkages. In Guyana, fieldwork took place from April to May 2012 and October to November 2012. During this time, I traveled the entire coastline across six coastal regions. I also had the opportunity to fly over the vast interior of the country to see the magnificent Kaieteur Falls, where the Potaro River drops from the immense edge of the Rupununi Savannah. The journey helped me to realize the depth of land that exists beyond the seawall. During the flight, I gazed down for hours at nothing but rainforest, an impenetrable thicket that was occasionally interrupted by one of the many mining operations taking place within. The remainder of my time in Guyana consisted of visits to coastal villages located within an hour's drive to the city center.

Throughout my fieldwork in Guyana, I listened to many different voices and viewpoints and observed both the community-based work of the EU-funded Guyana Mangrove Restoration Project and a series of small construction projects led by the Sea and River Defense Division of the Ministry of Public Works. My interviews were conducted in Georgetown,

Victoria, Berbice, Buxton, Mon Repot, Leguan, and Hope. I conducted a total of thirty in-depth interviews, mainly with community members, scientists, and government officials, mostly associated with seawall construction, mangrove restoration, or some combination of both. These were mainly one-on-one interviews that took place in a variety of locations including offices, homes, project sites, mangrove forests, the mud, and quite frequently on the seawall itself. During my stay in Guyana, I also spent time shadowing members of the GMRP, including mangrove forest rangers, coastal engineers, women producers, community officers, and project managers. Along the way, I ran into fishermen, school teachers, squatters, and curious elders, who happily sat down to talk with me and share their perspectives.

In the Maldives, fieldwork took place in April 2013 and again in March 2014, for a total of two months. Research was conducted across three major regions of the country, spanning the northernmost areas to the southernmost areas. My main point of contact was Mohamed Aslam, former minister of environment and housing in the Maldives and founder of the Land and Marine Environmental Research Center on the capital island of Malé, who connected me with key governmental figures, including the former president, Mohamed Nasheed. The majority of my interviews took place with people who reside near the seawall, including fishermen, residents, members of environmental organizations, scientists, and government officials.

I conducted a total of thirty-seven in-depth interviews with people in the Maldives who are involved in coastal protection or who live close to seawalls or eroding beaches. Due to language barriers, a few of the interviews were conducted in Dhivehi and later translated. Many of the historical documents used for this research were translated by Maldivian public intellectuals and published in an online archive for the purpose of promoting further scholarship in this area. Additional research was gathered through a variety of written sources, including published journals, ambassadorial notebooks, political documents, folklore, art, and social media. Given the politicized nature of historical representation in the country, perspectives varied widely along lines of religion, scientific expertise, and political ideology. This was taken into consideration, and documents and interviews were treated as discourses representative of larger narratives.

My interviews in Guyana and the Maldives addressed a range of questions pertaining to experiences and memories of the seawall. Generally, I was interested in how the development of seawalls and seawall alternatives involve and mobilize different groups of people, such as communities, development agencies, scientists, NGOs, and national and regional policy makers and donors. More specifically, I wanted to know the ways in which residents, coastal managers, engineers and politicians differed in their understandings and involvement with seawalls, as both agents of change and shoreline inhabitants. I interviewed public officials to understand issues associated with the location and funding of sea defense projects, but primarily focused on talking to ordinary people living along the seawall. Overall, my interviews took many shapes, as my informants differed greatly in terms of their experience and "expertise." Some lasted for more than two hours, and others were short and precise. I made it my objective to engage, to the best of my ability, with the particular and fascinating body of knowledge of my interviewees, whatever shape it should take.

Notes

INTRODUCTION: SEAWALL ENTANGLEMENTS

1. The decline of leatherback sea turtles has been linked to overfishing and overdevelopment and, more recently, to climate change. Warming temperatures are altering the sex determination of incubating hatchlings, resulting in an overabundance of females.

2. For example, see Nicola Dempsey, Harry Smith, and Mel Burton, eds., *Place-Keeping: Open Space Management in Practice* (New York: Routledge, 2014).

3. US Department of Arts and Culture, "Creative Placemaking, Placekeeping, and Cultural Strategies to Resist Displacement," USDAC Citizen Artist Salon, March 8, 2016, https://actionnetwork.org/forms/watch-the-creative-placekeeping-citizen-artist-salon.

4. *Resilience* finds its etymological origins and usage among Western thinkers such as Seneca the Elder, Pliny the Elder, Ovid, Cicero, and Livy (D. E. Alexander, "Resilience and Disaster Risk Reduction: An Etymological Journey," *Natural Hazards and Earth System Sciences* 13, no. 11 [2013]: 2707–16). The appeal of the term is particularly strong in the context of ecological crisis. The basic tenet of resilience thinking—that people and their built environments can return to some state of balance—has been adopted by climate action institutions such as the Intergovernmental Panel on Climate Change to shape policy. However, resilience has come under some scrutiny for its abstract nature (Juergen Weichselgartner and Ilan Kelman, "Geographies of Resilience: Challenges

and Opportunities of a Descriptive Concept," *Progress in Human Geography* 39, no. 3 [2015]: 249–67).

5. For a discussion of the difference between ecological and engineering metaphors of resilience, see Simon Davoudi, "Resilience: A Bridging Concept or a Dead End?," *Planning Theory & Practice* 13, no. 2 (2012): 299–333. Among the social sciences and in environmental planning circles there is a fixation with the engineering view of resilience, or "bounce-back-ability," which embraces resilience as the capacity to absorb shock and recover to a state of equilibrium. This is not so different from the ecological metaphor of equilibrium, which allows for multiple states of equilibrium and new equilibrium states.

6. Danny MacKinnon and Kate Driscoll Derickson, "From Resilience to Resourcefulness: A Critique of Resilience Policy and Activism," *Progress in Human Geography* 37, no. 2 (2012): 254.

7. For example, see Susan Fainstein, "Resilience and Justice," *International Journal of Urban and Regional Research* 39, no. 1 (2015): 157–67.

8. David Chandler, "Resilience and the End(s) of the Politics of Adaptation," *Resilience* 7, no. 3 (2019): 305.

9. Eija Meriläinen, "The Dual Discourse of Urban Resilience: Robust City and Self-Organised Neighbourhoods," *Disasters* 44, no. 1 (2020): 125–51.

10. Martin Fougère and Eija Meriläinen, "Exposing Three Dark Sides of Social Innovation through Critical Perspectives on Resilience," *Industry and Innovation* 28, no. 1 (2021): 1–18.

11. Malini Ranganathan and Eve Bratman, "From Urban Resilience to Abolitionist Climate Justice in Washington, DC," *Antipode* 53, no. 1 (2021): 115–37.

12. Kevin Grove, Savannah Cox, and Allain Barnett, "Racializing Resilience: Assemblage, Critique, and Contested Futures in Greater Miami Resilience Planning," *Annals of the American Association of Geographers* 110, no. 5 (2020): 1613–30.

13. R. Dean Hardy, Richard A. Milligan, and Nik Heynen, "Racial Coastal Formation: The Environmental Injustice of Colorblind Adaptation Planning for Sea-Level Rise," *Geoforum* 87 (2017): 62–72.

14. Greg Bankoff, "Remaking the World in Our Own Image: Vulnerability, Resilience and Adaptation as Historical Discourses," *Disasters* 43, no. 2 (2019): 221–39.

15. Helga Leitner et al., "Globalizing Urban Resilience," *Urban Geography* 39, no. 8 (2018): 1276–84.

16. Kasia Paprocki, "All That Is Solid Melts into the Bay: Anticipatory Ruination and Climate Change Adaptation," *Antipode* 51, no. 1 (2019): 295–315.

17. See Simon L. Lewis and Mark A. Maslin, "Defining the Anthropocene," *Nature* 519, no. 7542 (2015): 171–80. Despite the widespread acknowledgment that humans are changing the planet, there is not a strong consensus about which historical moment triggered the end of the Holocene and the beginning

of the Anthropocene, an epoch proposed by Paul Cruezten and Eugene Stoermer in the 1980s to account for observed human influences on the Earth system.

18. Jane Bennett, "The Agency of Assemblages and the North American Blackout," *Public Culture* 17 (2005), 445–66.

19. Andreas Malm, "Sea Wall Politics: Uneven and Combined Protection of the Nile Delta Coastline in the Face of Sea Level Rise," *Critical Sociology* 39, no. 6 (2013): 822.

20. Rosa Gonzales, "Community-Driven Climate Resilience Planning: A Framework, Version 2.0, "National Association of Climate Resilience Planners" (2017), https://movementstrategy.org/resources/community-driven-climate-resilience-planning-a-framework/.

21. Susanne Moser et al., "The Turbulent World of Resilience: Interpretations and Themes for Transdisciplinary Dialogue," *Climatic Change* 153, no. 1–2 (2019): 22.

22. Isabelle Anguelovski et al., "Equity Impacts of Urban Land Use Planning for Climate Adaptation: Critical Perspectives from the Global North and South," *Journal of Planning Education and Research* 36, no. 3 (2016): 333–48.

23. Thomas F. Gieryn, "A Space for Place in Sociology," *Annual Review of Sociology* 26, no. 1 (2000): 463–96.

24. Gieryn, "A Space for Place in Sociology," 471.

25. Visualizing and communicating this interconnection is the challenge of place-sensitive sociology. As Gieryn notes, "Place is, at once, the buildings, streets, monuments, and open spaces assembled at a certain spot *and* actors' interpretations, representations, and identifications. Both domains (the material and the interpretive, the physical and the semiotic) work autonomously *and* in a mutually dependent way" ("A Space for Place in Sociology," 466–67; original emphasis).

26. Rebecca Elliott, "The Sociology of Climate Change as a Sociology of Loss," *European Journal of Sociology* 59, no. 3 (2018): 301–37.

27. Elliott, "The Sociology of Climate Change," 307.

28. Elliott, "The Sociology of Climate Change," 310.

29. Elliott, "The Sociology of Climate Change," 312.

30. Anguelovski et al., "Equity Impacts of Urban Land Use Planning for Climate Adaptation."

31. As of 2022, it was estimated that 896 million people already live on low-lying coasts, and this number is expected to exceed one billion by 2050 (B. C. Glavovic et al., "Cross-Chapter Paper 2: Cities and Settlements by the Sea," in *Climate Change 2022: Impacts, Adaptation and Vulnerability. Contribution of Working Group II to the Sixth Assessment Report of the Intergovernmental Panel on Climate Change*, ed. H.-O. Pörtner et al. [Cambridge: Cambridge University Press, 2022], 2163–94).

32. David Matyas and Mark Pelling conceptualize the social and political dimensions of adaptive infrastructure on a continuum that differentiates interventions that maintain dominant power dynamics from those that push for a radical transformation of political systems (Matyas and Pelling, "Positioning Resilience for 2015: The Role of Resistance, Incremental Adjustment and Transformation in Disaster Risk Management Policy," *Disasters* 39, no. s1 [2014]: s14). In their analysis, seawalls fall into the first category of maintaining the status quo; however, my analysis of seawalls is more nuanced. Without attention to the underlying forces that make placekeeping a necessary burden for communities, seawalls are easily misunderstood.

33. See Nikhil Anand, Akhil Gupta, and Hannah Appel, eds., *The Promise of Infrastructure* (Durham, NC: Duke University Press, 2018).

34. Ash Amin, "Lively Infrastructure," *Theory, Culture & Society* 31, no. 7–8 (2014): 137.

35. For example, see Archie Davies, "The Coloniality of Infrastructure: Engineering, Landscape and Modernity in Recife," *Environment and Planning D: Society and Space* 39, no. 4 (2021): 740–57.

36. Octavia Butler, *Parable of the Sower* (New York: Warner Books, 1993).

37. Ursula K. Le Guin, *The Dispossessed* (New York: Harper & Row, 1974).

38. Marlen Haushofer, *The Wall*, trans. Shaun Whiteside (Pittsburgh, PA: Cleis Press, 1990).

39. Rachel Carson, *The Edge of the Sea* (New York: Houghton Mifflin, 1998), 189.

40. Carson, *The Edge of the Sea*, 189.

41. Carson, *The Edge of the Sea*, 189.

42. MacKinnon and Derickson, "From Resilience to Resourcefulness."

43. Alejandro De Coss-Corzo, "Patchwork: Repair Labor and the Logic of Infrastructure Adaptation in Mexico City," *Environment and Planning D: Society and Space* 39, no. 2 (2021): 238.

44. Kum-Kum Bhavnani, Peter Chua, and Dana Collins, "Critical Approaches to Qualitative Research," in *The Oxford Handbook of Qualitative Research*, ed. Patricia Leavy (New York: Oxford University Press, 2014), 176.

45. Michael Burawoy et al., eds., *Global Ethnography: Forces, Connections, and Imaginations in a Postmodern World* (Berkeley, CA: University of California Press, 2000).

46. Anna Lowenhaupt Tsing, *Friction: An Ethnography of Global Connection* (Princeton, NJ: Princeton University Press, 2004).

47. Jane Bennett, *Vibrant Matter: A Political Ecology of Things* (Durham, NC: Duke University Press, 2010).

48. Willson Harris, "A Note on the Genesis of the Guyana Quartet," in *The Guyana Quartet* (London: Faber and Faber, 1985), 7.

49. Philip Crowe, "The Maldive Islands," in *Diversions of a Diplomat in Ceylon*, ed. Philip K. Crowe (New York: D. Van Nostrand, 1956), 280–304.

CHAPTER 1. COASTAL DISRUPTION

1. "Mapping the World without Ice," *National Geographic* (Poster), September 2013. A follow-up piece printed in *The Atlantic* (November 2013) suggested that the map was actually too optimistic.

2. Based on insight from post-disaster landscapes, Weisman elegantly details the breakdown of taken-for-granted infrastructure—including the subway system of New York—and its reclamation by the wild (Alan Weisman, *The World Without Us* [New York: St. Martin's, 2007]).

3. While technological tools such as lidar and GIS have made possible the creation of highly detailed maps for assessing the threat of sea change, these models, no matter how beautifully rendered or thoughtfully accounted for, cannot capture all the trace details that together compose the living world; wind patterns, storm surges, shifting ocean currents, and the piecemeal construction of a seawall are necessarily left unresolved. Sea change is not an absolute measure with a uniform horizontal surface, like water in a bathtub. Its flows are incongruous, and the land moves as well, sinking and shifting with the forces of intensified land use and the natural forces of the great crustal plates. In many cases, sea level rise is mistaken for land subsidence.

4. Langdon Winner, "Do Artifacts Have Politics?," *Daedalus* 109, no. 1 (1980): 121–36.

5. Bernward Joerges challenges Langdon Winner's account of the low-lying bridges in order to point to what he calls "the seductive power of parables" and suggests that we might conceive of infrastructure as objects that "serve as media or mediation, negotiation and translation between the reciprocal expectations and requirements of many people or organizations" (Bernward Joerges, "Do Politics Have Artefacts?," *Social Studies of Science* 29, no. 3 [1999]: 424).

6. Malm, "Sea Wall Politics."

7. Susan Leigh Star and James R. Griesemer, "Institutional Ecology, 'Translations' and Boundary Objects: Amateurs and Professionals in Berkeley's Museum of Vertebrate Zoology, 1907–39," *Social Studies of Science* 19, no. 3 (1989): 387–420.

8. Isto Huvila, "The Politics of Boundary Objects: Hegemonic Interventions and the Making of a Document," *Journal of the American Society for Information Science and Technology* 62, no. 12 (2011): 2528–39.

9. Wolfgang Sachs, ed., *The Development Dictionary: A Guide to Knowledge as Power*, 2nd ed. (New York: Zed Books, 2010), 122.

10. Bankoff, "Remaking the World in Our Own Image," 222.

11. Theodor W. Adorno and Max Horkheimer, *Dialectic of Enlightenment*, trans. Edmund Jephcott, ed. Gunzelin Schmid Noerr (Stanford, CA: Stanford University Press, 2002).

12. Maladaptation is highlighted as a concern in the 2022 International Panel on Climate Change (IPCC) report, defined as actions that "can create lock-ins

of vulnerability, exposure and risks that are difficult and expensive to change and exacerbate existing inequalities" (IPCC, "Summary for Policymakers," in H.-O. Pörtner et al., *Climate Change 2022*, 29).

13. Naomi Klein, *The Shock Doctrine: The Rise of Disaster Capitalism* (New York: Picador, 2007).

14. Tony Bennett et al., eds., *New Keywords: A Revised Vocabulary of Culture and Society* (Malden, MA: Blackwell, 2005), 43.

15. Lewis and Maslin, "Defining the Anthropocene."

16. Lewis and Maslin, "Defining the Anthropocene," 174.

17. Rendel Palmer, "History of Coastal Engineering in Great Britain," in *History and Heritage of Coastal Engineering: A Collection of Papers on the History of Coastal Engineering in Countries Hosting the International Coastal Engineering Conference 1950–1996*, ed. Nicholas C. Kraus (New York: American Society of Civil Engineers, 1996), 238.

18. Michael R. Gourlay, "History of Coastal Engineering in Australia," in Kraus, *History and Heritage of Coastal Engineering*, 6.

19. Atsuro Morita, "Infrastructuring Amphibious Space: The Interplay of Aquatic and Terrestrial Infrastructures in the Chao Phraya Delta in Thailand," *Science as Culture* 25, no. 1 (2016): 128.

20. Sachs, *The Development Dictionary*, x.

21. Arturo Escobar, "Planning," in Sachs, *The Development Dictionary*," 145–60.

22. In March 2021, the UN formally adopted an ecosystem accounting framework to replace the conventional GDP model, placing a monetary value on the natural environment.

23. David Sogge, "Multilateral Actors in Development," in *Introduction to International Development: Approaches, Actors, and Issues*, ed. Alexander Paul Haslam, Jessica Schafer, and Pierre Beaudet (Oxford: Oxford University Press, 2009), 170.

24. Philip McMichael, *Development and Social Change: A Global Perspective* (Los Angeles: Sage, 2017), 46.

25. Uma Kothari, "Authority and Expertise: The Professionalisation of International Development and the Ordering of Dissent," *Antipode* 37, no. 3 (2005): 425–46.

26. Kothari, "Authority and Expertise," 434.

27. Alvin Y. So, *Social Change and Development: Modernization, Dependency, and World-Systems Theories* (Newbury Park, CA: Sage, 1990), 23–33.

28. Walt Whitman Rostow, *The Stages of Economic Growth: A Non-Communist Manifesto* (Cambridge: Cambridge University Press, 1960).

29. So, *Social Change and Development*, 53.

30. The earliest usage of the term *development* conveyed a notion of evolutionary processes and was used to describe and explain natural growth patterns

in plants and animals (Gustavo Esteva, "Development," in Sachs, *The Development Dictionary*, 1–23). As such, development came to connote the evolution of species and the realization of the innate potentialities of living things. In the eighteenth and nineteenth centuries, the terms *development* and *evolution* became coterminous, used interchangeably by scientists. By the late eighteenth century, the biological metaphor of development was carried into the social sphere to explain historical phenomena. During this time, development came to be understood philosophically as "social progress" and was widely embraced in the political space. The evolution of societies to an appropriate form, from the status of "traditional" to the status of "modern," became a widely adopted perspective. The early notion of social progress was closely linked to the necessity of reason and science, and the earliest applications of development thinking were manifest in projects of social engineering, including projects of colonization (Gilbert Rist, *The History of Development: From Western Origins to Global Faith*, 2nd ed. [London: Zed Books, 2002]).

31. Seymour Martin Lipset, *Political Man: The Social Bases of Politics* (New York: Doubleday, 1960).

32. So, *Social Change and Development*, 36.

33. McMichael, *Development and Social Change*, 110.

34. Robert O. Keohane and Joseph S. Nye, "Globalization: What's New? What's Not? (And So What?)," *Foreign Policy* 118 (2000): 104–19.

35. McMichael, *Development and Social Change.*

36. Naomi Klein, *The Battle for Paradise: Puerto Rico Takes on Disaster Capitalists* (Chicago: Haymarket, 2018).

37. Torben Sørensen, Jorgen Fredsøe, and Per Roed Jakobsen, "History of Coastal Engineering in Denmark," in Kraus, *History and Heritage of Coastal Engineering*, 103–41.

38. Hanz D. Niemeyer, Hartmut Eiben, and Hans Rohde, "History and Heritage of German Coastal Engineering," in Kraus, *History and Heritage of Coastal Engineering*, 169–71.

39. Palmer, "History of Coastal Engineering in Great Britain," 219.

40. Ann Laura Stoler, ed., *Imperial Debris: On Ruins and Ruination* (Durham, NC: Duke University Press, 2013), 23.

41. Catia Antunes, "Early Modern Ports, 1500–1750," *European History Online* (*EGO*) (2010), https://d-nb.info/103126227X/34.

42. Niemeyer, Eiben, and Rohde, "History and Heritage of German Coastal Engineering," 199.

43. Chandra Mukerji, "Stewardship Politics and the Control of Wild Weather: Levees, Seawalls, and State Building in 17th-Century France," *Social Studies of Science* 37 (2007): 127–33.

44. Mukerji, "Stewardship Politics and the Control of Wild Weather," 130.

45. Brian Larkin, *Signal and Noise: Media, Infrastructure, and Urban Culture in Nigeria* (Durham, NC: Duke University Press, 2008), 47.

46. The terrifying and sublime magnitude of water-based infrastructure is captured in *Watermark*, a 2012 documentary film by Jennifer Baichwal and Edward Burtynsky. Across a vast and entangled assemblage of infrastructure and cultural symbols, *Watermark* transports the viewer into the stunning world of human engagements with water, from the building of massive aquaculture farms to hydroelectric dams.

47. The damage was not limited to the colonies but was also felt in the heart of empire. With the rise of industrialization, Venice nearly came to an end. Unregulated and booming with new technologies, industries on the mainland pumped enough water from the area's underground reservoirs that Venice sank at a rate nearly twice that of previous centuries. Eventually, in the 1970s, awareness put an end to the pumping, but the damage was done and irreversible. In addition to the human-induced sinking of Venice, there was the convergence of human-induced degradation of the lagoon, which caused an increase in sea level as protective barriers deteriorated. The salt marshes and sandbanks that were once under legal protection in Venice began to die off and nearly disappear.

48. Industrial sand mining was simultaneously undermining the natural barriers and integrity of coastlines. In Australia, a conflict emerged in the mid-twentieth century over the extraction of sediments found on the beaches and dunes of New South Wales and Queensland to provide titanium to the United States aerospace industry (Gourlay, "History of Coastal Engineering in Australia," 11). In Great Britain, the Coastal Protection Act was initiated in 1949 to restrict the removal of materials from the beaches: directly in response to growing concerns over beach erosion (Palmer, "History of Coastal Engineering in Great Britain," 227).

49. Kraus, *History and Heritage of Coastal Engineering*.

50. Kraus, *History and Heritage of Coastal Engineering*, iii.

51. Johan van Veen, *Dredge Drain Reclaim*, 4th ed. (Dordrecht: Springer, 1955), 158.

52. Van Veen, *Dredge Drain Reclaim*, 174.

53. The rapid industrial development of the coast also coincides with the emergence of the military-industrial-science complex and the privatization of coastal engineering research. In Britain, the privatization of national hydraulics laboratories for commercial purposes post–World War II raised concerns about the funding of basic research (Palmer, "History of Coastal Engineering in Great Britain").

54. In Great Britain, seawalls were used in the late nineteenth century to prevent flooding and to stabilize eroding cliffs, which led to the development of seaside tourist towns (Palmer, "History of Coastal Engineering in Great Britain"). In Germany, the island of Norderney was established as an official holiday resort

in the late sixteenth century, increasing property values (Niemeyer Eiben, and Rohde, "History and Heritage of German Coastal Engineering," 185).

55. Gourlay, "History of Coastal Engineering in Australia," 11.

56. Luc Hamm, "History of Coastal Engineering in France," in Kraus, *History and Heritage of Coastal Engineering*, 159.

57. Antonio J. Maza, Rodolfo Silva, and Carlos Sanchez, "History of Coastal Engineering in Mexico," in Kraus, *History and Heritage of Coastal Engineering*, 387.

58. Palmer, "History of Coastal Engineering in Great Britain," 235.

59. CIRIA, CUR, and CETMEF, *The Rock Manual: The Use of Rock in Hydraulic Engineering*, 2nd ed. (London: CIRIA, 2007), 71.

60. CIRIA, CUR, and CETMEF, *The Rock Manual*, 86.

61. CIRIA, CUR, and CETMEF, *The Rock Manual*, 248.

62. CIRIA, CUR, and CETMEF, *The Rock Manual*, 260.

63. For example, see David Naguib Pellow, *Resisting Global Toxics: Transnational Movements for Environmental Justice* (Cambridge, MA: MIT Press, 2007).

64. See Stephen Zavestoski, "The Struggle for Justice in Bhopal: A New/Old Breed of Transnational Social Movement," *Global Social Policy* 9, no. 3 (2009): 383–407; and Steven Donzinger, Laura Garr, and Aaron Marr Page, "Rainforest Chernobyl Revisited: The Clash of Human Rights and BIT Investor Claims: Chevron's Abusive Litigation in Ecuador's Amazon," *Human Rights Brief* 17, no. 2 (2010): 1–8.

65. James O'Conner, "Is Sustainable Capitalism Possible?," in *Is Capitalism Sustainable*, ed. Martin O'Conner (New York: Guilford Press, 1994), 152–75.

66. John Bellamy Foster, "Capitalism and Ecology: The Nature of the Contradiction" (Paper presented at the Socialism Conference, Chicago, 2002).

67. Sachs, *The Development Dictionary*, 34.

68. WCED, *Our Common Future* (Oxford: Oxford University Press, 1987), 3.

69. Peter Dicken, *Global Shift: Mapping the Changing Contours of the World Economy*, 5th ed. (London: Sage, 2007), 544.

70. Esteva, "Development," 13.

71. Andre Gunder Frank, "The Development of Underdevelopment," *Monthly Review* 18, no. 4 (1966): 17–31.

72. This form of scientific forestry finds its origins in Prussia and Saxony during the late eighteenth century; the quantification of land was a process by which states could simplify a terrain, its products, and its people to make them more legible for state officials to manage. See James C. Scott, *Seeing Like a State: How Certain Schemes to Improve the Human Condition Have Failed* (New Haven, CT: Yale University Press, 1999).

73. Mark Dowie, *Conservation Refugees: The Hundred-Year Conflict Between Global Conservation and Native Peoples* (Cambridge, MA: MIT Press, 2009).

74. Vandana Shiva, *Staying Alive: Women, Ecology and Development* (London: Zed Books, 1989).

75. Rist, *The History of Development.*

76. Vandana Shiva, "Ecological Balance in an Era of Globalization," in *The Globalization Reader*, ed. Frank J. Lechner and John Boli (London: Routledge, 2000), 567.

77. For example, see Greg Browder et al., *Integrating Green and Gray: Creating Next Generation Infrastructure* (Washington, DC: World Bank and World Resources Institute, 2019).

78. According to one estimate: "Globally, about 33,700 km^2 of land has been gained from the sea during the last 30 years (about 50% more than has been lost), with the biggest gains due to land reclamation in places like Dubai, Singapore and China" (Alexander Bisaro et al., "Leveraging Public Adaptation Finance through Urban Land Reclamation: Cases from Germany, the Netherlands and the Maldives," *Climatic Change* 160, no. 4 [2020]: 672).

79. Mark Jackson and Veronica della Dora, "'Dreams So Big Only the Sea Can Hold Them': Man-Made Islands as Anxious Spaces, Cultural Icons, and Travelling Visions," *Environment and Planning A: Economy and Space* 41 (2009): 2086–104.

80. Jackson and della Dora, "'Dreams So Big Only the Sea Can Hold Them,'" 2091.

81. Jackson and della Dora, "'Dreams So Big Only the Sea Can Hold Them,'" 2101.

82. In Britain, for example, the Campaign for a Living Coast emerged in the early 1990s. See Palmer, "History of Coastal Engineering in Great Britain," 248.

83. Emma Marris, *Rambunctious Garden: Saving Nature in a Post-Wild World* (New York: Bloomsbury, 2011).

84. Joshua J. Cousins, "Justice in Nature-Based Solutions: Research and Pathways," *Ecological Economics* 180 (2021), https://doi.org/10.1016/j.ecolecon.2020.106874.

85. L. Rist et al., "Applying Resilience Thinking to Production Ecosystems," *Ecosphere* 5, no. 6 (2014), http://dx.doi.org/10.1890/ES13-00330.1.

86. Chandler, "Resilience and the End(s) of the Politics of Adaptation."

87. See also Julie Sze, *Fantasy Islands: Chinese Dreams and Ecological Fears in an Age of Climate Crisis* (Berkeley: University of California Press, 2015).

88. In some areas, old land reclamation projects are in the process of being undone. Managed realignment in Britain, for example, involves a process of undoing historical land reclamation for biodiversity and flood defense that repositions actively maintained flood defense lines landward in areas that were historically reclaimed for agriculture.

89. For example, the People's Coalition for Fisheries Justice Indonesia and the Indonesian Traditional Fishermen's Association argue urgently against the "Giant Sea Wall" planned for a newly reclaimed area of Jakarta, which would displace

thousands of local people and fishermen to accommodate a project that would be ineffective in resolving the flooding and water crisis that has long disrupted the lives of residents. See Thanti Octavianti and Katrina Charles, "Disaster Capitalism? Examining the Politicisation of Land Subsidence Crisis in Pushing Jakarta's Seawall Megaproject," *Water Alternatives* 11, no. 2 (2018): 394–420.

90. William Cronon, "The Trouble with Wilderness; Or, Getting Back to the Wrong Nature," *Environmental History* 1, no. 1 (1996): 7–28.

91. For example, see Palmer, "History of Coastal Engineering in Great Britain," 248.

92. Benjamin Kline, *First along the River: A Brief History of the US Environmental Movement* (New York: Arcada Books, 2000).

93. Ramachandra Guha and Juan Martinez-Alier maintain that environmentalism is not a static category but one that entails relationships of power and competing ideologies; environmentalism, they argue, is "red on the outside, but green on the inside" (Guha and Martinez-Alier, *Varieties of Environmentalism: Essays North and South* [London: Earthscan, 2006], 16).

94. Sachs, *The Development Dictionary*, 124.

95. Chandler, "Resilience and the End(s) of the Politics of Adaptation," 308.

96. Anna Tsing, "The Buck, the Bull, and the Dream of the Stag: Some Unexpected Weeds of the Anthropocene," *Suomen Antropologi: Journal of the Finnish Anthropological Society* 42, no. 1 (2017): 3–21.

97. Isabelle Anguelovski et al., "Why Green 'Climate Gentrification' Threatens Poor and Vulnerable Populations," *Proceedings of the National Academy of Sciences of the United States of America* 116, no. 52 (2019): 26139–43.

CHAPTER 2. THE STRANGLED SHORE

1. Ray Kril and Rupert Roopnaraine, dir., *The Seawall, Tales from the Guyana Coast* (Guyana, 1992).

2. Henry G. Dalton, *The History of British Guiana*, vol. 1 (London: Longman, Brown, Green, and Longmans, 1855), 3.

3. High spring tides occur when the sun and moon are nearly in alignment with Earth, resulting in an increased gravitational pull.

4. It is unlikely that the Dutch were unaware of the mobility of the coastal plain, as they were familiar with such dynamics in their own country. Perhaps they did not intend to stay longer than what was necessary for capitalizing on the fertile ground pulled from the intertidal zone.

5. Allison A. Mead and Michael T. Lee, "Sediment Exchange between Amazon Mudbanks and Shore-Fringing Mangroves in British Guiana," *International Journal of Marine Geology, Geochemistry and Geophysics* 208 (2004): 170.

6. Johan van Veen, *Dredge Drain Reclaim: The Art of a Nation*, 4th ed. (Dordrecht: Springer, 1955). First published in German in 1948 and translated into English in 1952. Van Veen was the chief coastal engineer of the Netherlands, put out of work by World War II. He took to writing about the historical making of the Netherlands, starting with the early Dutch settlers. He is credited with inspiring the construction of the Delta Works.

7. Van Veen, *Dredge Drain Reclaim*, 15. The early people of the Netherlands struggled to find space and autonomy, pushed to precarity by repression and isolation, faced with the burden of reclaiming land from the sea.

8. Pliny the Elder, "Countries That Have No Trees," in *Natural History* (London: Taylor and Francis, 1855).

9. Niemeyer, Eiben, and Rohde, "History and Heritage of German Coastal Engineering," 183.

10. Van Veen, *Dredge Drain Reclaim*, 24.

11. Van Veen, *Dredge Drain Reclaim*, 26.

12. Van Veen, *Dredge Drain Reclaim*, 24.

13. Van Veen, *Dredge Drain Reclaim*, 27.

14. Van Veen, *Dredge Drain Reclaim*, 30.

15. Van Veen, *Dredge Drain Reclaim*, 80.

16. Adam Sundberg, "Molluscan Explosion: The Dutch Shipworm Epidemic of the 1730s," *Arcadia: Explorations in Environmental History*, no. 14 (2015), https://doi.org/10.5282/rcc/7307.

17. The Dutch tried covering the dikes with tar, lining the bottoms with copper, and laying stones at the base of the pilings. Nevertheless, *T. navalis* flourished, eroding the sea defenses. Ultimately, the Dutch won the battle against the sea worms but not without a drastic change to the landscape.

18. Geo D. Bayley, *Handbook of British Guiana, 1909* (London: Forgotten Books, 2018).

19. Louis Stanislas d'Arcy de La Rochette, *Guyana Coast* (London: William Faden, 1783).

20. Vere T. Daly, *The Making of Guyana* (London: Macmillan, 1974).

21. Daly, *The Making of Guyana*, 73.

22. Walter Rodney, *A History of the Guyanese Working People, 1881–1905* (Baltimore: Johns Hopkins University Press, 1981), 2.

23. Daly, *The Making of Guyana*, 73.

24. Van Veen, *Dredge Drain Reclaim*, 15.

25. Rodney, *A History of the Guyanese Working People*, 2.

26. Rodney, *A History of the Guyanese Working People*, 6.

27. Rodney, *A History of the Guyanese Working People*, 90–91.

28. Walter Rodney, *Guyanese Sugar Plantations in the Late Nineteenth Century: A Contemporary Description from the Argosy* (Georgetown, Guyana: Release Publishers, 1979), xiv.

29. For example, see "Maps of the Orinoco-Essequibo Region," commissioned by the United States–Venezuela–British Guiana Boundary Commission in 1897.

30. Rodney, *A History of the Guyanese Working People.* British colonial land records mention areas measured in acres, roods, and poles, where a rood is equivalent to 1,210 square yards.

31. Rodney, *A History of the Guyanese Working People*, 5.

32. Philip Edmond Wodehouse, "Correspondence Respecting the Sea Wall at Demerara," in *Accounts and Papers of the House of Commons 1854–55* (London: House of Commons).

33. James Rodway and James Graham Cruickshank, *The Story of Georgetown: Revised from a Series of Articles in the "Argosy" 1903* (Demerara, Guyana: Argosy, 1920).

34. Rodway and Cruickshank, *The Story of Georgetown.*

35. Rodney, *A History of the Guyanese Working People*, 15–16.

36. Rodney, *A History of the Guyanese Working People*, 12.

37. *Groin* is the US spelling, and *groyne* is the British spelling. Some of the quoted passages use the British spelling.

38. National Development Strategy Secretariat, "Water Management and Flood Control," http://guyana.org/NDS/chap40.htm#2contents_A.

39. For more on the history of land reclamation in Great Britain, see Michael Williams, *The Draining of the Somerset Levels* (Cambridge: Cambridge University Press, 1970).

40. Edward Case, "The Dymchurch Wall and Reclamation of Romney Marsh," *The Engineer* 88 (1899), 407–8.

41. Case, "The Dymchurch Wall," 408.

42. Gerald O. Case, *Coast Erosion Protection Works on the Case System in British Guiana* (New York: Tidal Engineering Corporation, 1920), 5.

43. "Obituary. Edward Case, 1842–1899," *Minutes and Proceedings of the Institution of Civil Engineers* 139 (1900): 374–76.

44. "Past Evolution of Politics: Sea Defences," Government of Guyana, 1996, accessed August 8, 2020, http://www.guyana.org/NDS/chap40.htm#2contents_A.

45. Case, *Coast Erosion Protection Works*, 30.

46. Case, *Coast Erosion Protection Works*, 5.

47. Gerald Otley Case, Patent US20254938A: Reinforcement of Concrete or the Like, 1940.

48. Case, *Coast Erosion Protection Works*, 15.

49. Case, *Coast Erosion Protection Works*, 17.

50. Case, *Coast Erosion Protection Works*, 15.

51. "Past Evolution of Politics: Sea Defences."

52. Lloyd F. Kandasammy, "A Brief History of Floods in Guyana," *Stabroek News* (Guyana), February 6, 2006, http://www.landofsixpeoples.com/news601/ns6021650.htm.

53. Kandasammy, "A Brief History of Floods in Guyana."

54. Palmer, "History of Coastal Engineering in Great Britain," 238.

55. Rodney, *A History of the Guyanese Working People*, 103.

56. Rodney, *A History of the Guyanese Working People*, 2.

57. Rodney, *A History of the Guyanese Working People*, 9.

58. Arthur Lewis, *Labour in the West Indies: The Birth of a Workers' Movement* (London: Fabian Society, 1939); quoted in Cheddi Jagan, *Fight for Freedom: Waddington Constitution Exposed* (Georgetown, Guyana: C. Jagan, 1952).

59. Jagan, *Fight for Freedom*.

60. Jagan, *Fight for Freedom*.

61. "The Suspension of the British Guiana Constitution—1953 (Declassified British Documents)," updated August 2004, accessed August 8, 2020, http://www.guyana.org/govt/declassified_british_documents_1953.html.

62. Cheddi Jagan, "Death to Imperialism," Cheddi Jagan Research Center, 1954, accessed August 8, 2020, https://jagan.org/CJ%20Articles/Early%20Aricles/early_cj.html.

63. Janet Jagan, "A Piece of Guyana's History, 1953–1955," *Thunder: Theoretical and Analytical Journal of the People's Progressive Party of Guyana* 15, no. 1 (1983), available at https://jagan.org/Janet%20Jagan/JJ%20Articles/images/7259.pdf.

64. For more on the involvement of the United States in Burnham's rise to power, see Stephan G. Rabe, *U.S. Intervention in British Guiana: A Cold War Story* (Chapel Hill: University of North Carolina Press, 2005); and Shanya Cordis, "In Guyana, Colonial Regimes Power the New Oil Frontier," *NACLA Report on the Americas* 53, no. 3 (2021): 268–74.

65. Odeen Ishmael, *The Guyana Story: From Earliest Times to Independence* (Bloomington, IN: Xlibris, 2013).

66. US Department of State, "Progress Report to the 303 on Guyana" by Richard Lehman, Washington DC, June 17, 1969, https://static.history.state.gov/frus/frus1969-76ve10/pdf/d366.pdf.

67. World Bank, "Appraisal Report: Sea Defense Project Guyana, International Bank for Reconstruction and Development" (Washington, DC, 1968), ii, http://documents1.worldbank.org/curated/en/291571468275676329/text/multiopage.txt.

68. According to the report, "Sling-mud is one of the most remarkable phenomena of the Guyana coast. It is formed by the orbital movement of water particles stirring up the partly liquid, partly plastic mass of mud resting on the seabed. The mud particles are moved, consuming so much energy that the waves are damped out. Areas of sling-mud can be readily spotted from the air and their presence is easily detected from small boats. Their arrival usually heralds the beginning of accretion" (World Bank, "Appraisal Report: Sea Defense Project Guyana," 2).

69. World Bank, "Appraisal Report: Sea Defense Project Guyana," 7.

70. National Development Strategy Secretariat, "Water Management and Flood Control," Ministry of Finance, Guyana, 1996, http://guyana.org/NDS/chap40.htm#2contents_A.

71. World Bank, "Appraisal of Second Sea Defense Project (Georgetown Urban Protective Works) Guyana," (Washington, DC, 1971), http://documents1.worldbank.org/curated/en/574601468250853198/text/multi-page.txt.

72. Ashton Chase, *Guyana—a Nation in Transit: Burnham's Role* (Georgetown, Guyana: Paunik, 1994); quoted in Richard L. Cheltenham, Seenath Jairam, and Jacqueline Samuels-Browne, "Report of the Commission of Inquiry Appointed to Enquire and Report on the Circumstances Surrounding the Death of the Late Dr. Walter Rodney on Thirteenth Day of June, One Thousand Nine Hundred and Eighty at Georgetown (Parliament of the Co-operative Republic of Guyana, 2016)," http://parliament.gov.gy/documents/documents-laid/.

73. Walter Rodney, "The Struggle Goes On: A Speech Made in September 1979" (History Is a Weapon: Working People's Alliance, 1984), https://www.historyisaweapon.com/defcon1/rodnstrugoe.html.

74. Cheltenham, Jairam, and Samuels-Browne, "Report of the Commission of Inquiry."

75. Cheltenham, Jairam, and Samuels-Browne, "Report of the Commission of Inquiry."

76. "The Georgetown Sea Walls Then and Now" (Real Guyana), Facebook, October 5, 2013, https://www.facebook.com/RealGuyana/photos/a.531574053579372/531575656912545.

77. "The Georgetown Sea Walls Then and Now."

78. Stuart Hall, "Cultural Identity and Diaspora," *Framework: The Journal of Cinema and Media* 36 (1989): 234.

79. Hall, "Cultural Identity and Diaspora," 235.

80. Van Veen, *Dredge Drain Reclaim*, 174.

CHAPTER 3. LOST ORIGINS: DREAMS OF A GREEN SEAWALL

1. The project was conceived by Witness Project and a French street artist who uses his camera "to turn the world inside out." Under the slogan, "What we see changes who we are," the eyes were fashioned as a response to the mounting problem of gendered violence in the country. The photographs, collected by the children themselves, were prompted by questions like, "How do you look when you feel scared?" The artist, who goes by JR, delivered TED talks in March 2011 and May 2012. The Guyana seawall project was documented in 2012 on the Facebook page of Witness Project.

2. The public discourse on climate change in Guyana was almost nonexistent when I conducted my fieldwork there in 2012. This perhaps has more to

do with the politicized nature of the discourse than a calculated disbelief in science.

3. Coss-Corzo, "Patchwork," 238.

4. As stated on the website of the Ministry of Public Works on September 6, 2013, https://mopw.gov.gy/tags/seawall.

5. Carson, *The Edge of the Sea*, 1.

6. Carson, *The Edge of the Sea*, 2.

7. Carson, *The Edge of the Sea*, 1.

8. Laura A. Ogden, *Swamplife: People, Gators, and Mangroves Entangled in the Everglades* (Minneapolis: University of Minnesota Press, 2011).

9. Elizabeth Rush, *Rising: Dispatches from the New American Shore* (Minneapolis, MN: Milkweed Editions, 2018).

10. Environmental Protection Agency and Ministry of Natural Resources and the Environment Georgetown, "Fifth National Report to the Convention on Biological Diversity," September 2014, https://www.cbd.int/doc/world/gy/gy-nr-05-en.pdf.

11. Orrin H. Pilkey and Rob Young, *The Rising Sea* (Washington, DC: Island Press, 2009).

12. Edward J. Anthony and Nicolas Gratiot, "Coastal Engineering and Large-Scale Mangrove Destruction in Guyana, South America: Averting an Environmental Catastrophe in the Making," *Ecological Engineering* 47 (2012): 268–73.

13. Annette Arjoon-Martins's backstory has been covered by numerous sources. According to one source, Arjoon-Martins became involved in environmental advocacy and coastal protection in the early 2000s. During this time, she was hired to guide a scientist to the white sandy shores of Shell Beach on the far western edge of Guyana's coastline, a territory in dispute between Guyana and Venezuela. She discovered endangered sea turtles nesting on the beach. At this moment, she found her path. Over the next few years, Annette worked to enact a policy that would make Shell Beach Guyana's first and only protected coastal area. See Cedriann J. Martin, "Annette Arjoon-Martins: 'The Grassroots People Are the Real Heroes,'" *Caribbean Beat*, 2011, https://www.caribbean-beat.com/issue-110/-grassroots-people-are-real-heroes#axzz7Z8KE1zNU.

14. For more on the epistemological nuances and experimental nature of scientific modeling to determine favorable planting sites, see Sarah E. Vaughn, "Disappearing Mangroves: The Epistemic Politics of Climate Adaptation in Guyana," *Cultural Anthropology* 32, no. 2 (2017): 242–68.

15. Andrea Martinez, "Gender and Development: Issues and Struggles of Third World Women," in *Introduction to International Development: Approaches, Actors, and Issues*, ed. Paul A. Haslam, Jessica Schafer, and Pierre Beaudet (Oxford: Oxford University Press, 2009), 83.

16. Jill M. Belsky, "Gender, Community, and the Politics of Community-Based Rural Ecotourism in Belize," in *Contested Nature: Promoting International*

Biodiversity Conservation with Social Justice in the Twenty-First Century, ed. Steven R. Brechin et al. (New York: State University of New York Press, 2003), 92.

17. Peter R. Wilshusen, "Exploring the Political Contours of Conservation," in Brechin et al., *Contested Nature*, 41–58.

18. Arun Agrawal and Clark C. Gibson, "Enchantment and Disenchantment: The Role of Community in Natural Resource Conservation," *World Development* 27, no. 4 (1999): xii. Some argue that community-based projects should begin by asking fundamental questions about who will participate in and benefit from decisions about how conservation will occur and at what social cost (Brechen et al., *Contested Nature*). Answers to such questions must take into account issues of power as articulated through the uneven impacts of conservation programs on a diverse array of people to avoid further disenfranchisement.

19. See Anthony and Gratiot, "Coastal Engineering," 271–72, where they write, "At present, the cost of coastal protection in Guyana works out at US$2000 (€1600) for a 100 m-long compacted earth dike, US$350,000 (€280,000) for a 100 m-long rock armour dike, and US$5000–20,000 (€4000–16,000) for a 100 m × 20 m swathe of replanted mangroves."

20. European Commission, Directorate-General for International Cooperation and Development, *Using Innovative and Effective Approaches to Deliver Climate Change Support to Developing Countries* (2012), https://data.europa.eu/doi/10.2783/76697.

21. Anthony and Gratiot, "Coastal Engineering," 272.

22. Dominick A. DellaSala, "'Real' vs. 'Fake' Forests: Why Tree Plantations Are Not Forests," in *Encyclopedia of the World's Biomes*, ed. Michael I. Goldstein and Dominick A. DellaSala (Oxford: Elsevier, 2020), 47–55.

23. Hall, "Cultural Identity and Diaspora."

24. Hall, "Cultural Identity and Diaspora," 236.

25. Marris, *Rambunctious Garden*, 3. The concept of shifting baselines describes how our perceptions are influenced by the creeping rate of change around us, affecting what counts as "normal" (Daniel Pauly, "Anecdotes and the Shifting Baseline Syndrome of Fisheries," *Trends in Ecology & Evolution* 10, no. 10 [1995]: 430). Such baselines are embedded everywhere in our infrastructural lives, foregrounding and shaping what is considered ordinary and desirable.

26. Mélanie Gidel, "Fragmentation on the Waterfront: Coastal Squatting Settlements and Urban Renewal Projects in the Caribbean," in *Transforming Urban Waterfronts: Fixity and Flow*, ed. Gene Desfor et al. (New York: Routledge, 2011), 35–53.

27. Gidel, "Fragmentation on the Waterfront," 40.

28. One of his songs from the 1970s, "Not a Blade of Grass," launched him into the political spotlight. The song was a response to the Venezuela-Guyana border conflict. In thinking about this dispute, Dave drew on the words of Chief Seattle from the American West, who, he says, "would not give up one river, not one

buffalo, not one valley, not even one blade of grass to the invading Whites" (Guyanese Online, "Dave Martins—the Story Behind 'Not a Blade of Grass,'" https://guyaneseonline.net/2012/08/03/dave-martins-the-story-behind-not-a-blade-of-grass/).

29. Michael Jordan, "Dave Martins Guyanese Cultural Icon," *Kaieteur News*, May 9 2010, https://www.kaieteurnewsonline.com/2010/05/09/dave-martins-guyanese-cultural-icon/.

30. Dave Martins, "Vreed en Hoop Then and Now," *Stabroek News* (Guyana), August 23, 2010, https://www.stabroeknews.com/2010/08/23/features/vreed-en-hoop-then-and-now/.

31. Martins, "Vreed en Hoop Then and Now."

32. Amitav Ghosh, *The Hungry Tide* (New York: Mariner Books, 2006), 50.

33. Anna J. Wesselink et al., "Dutch Dealings with the Delta," *Nature and Culture* 2, no. 2 (2007): 202.

34. See L. Rist et al., "Applying Resilience Thinking to Production Ecosystems."

35. Marris, *Rambunctious Garden*, 125.

36. From a 2007 interview, "How to Work with Nature by Henk Ovink," in *Planificando Arquitectura—Planning Architecture* (2017), https://www.youtube.com/watch?v=HWOTDeq4qoc&feature=youtu.be.

37. Ovink claims that farmers were part of the planning process and thus had a say in creating their farm buildings on high parts of the land so their cattle could stay.

38. Wesselink et al., "Dutch Dealings with the Delta."

39. While this claim is true, it can be misleading. The assumption that beekeeping will increase mangrove populations is contested and driven by the underlying tensions of expert knowledge (Vaughn "Disappearing Mangroves, 258–61).

40. Shiva, *Staying Alive*.

41. For some ecofeminists, environmental degradation is also seen as an outcome of capitalist relations, deeply rooted in the patriarchal forms of subjugation applied to both "nature" and women (see, e.g., Sherry B. Ortner, "Is Female to Male as Nature Is to Culture?," in *Women, Culture, and Society*, ed. M. Z. Rosaldo and L. Lamphere [Stanford, CA: Stanford University Press, 1974], 68–87). However, as the Indian development economist Bina Agarwal argues, ecofeminism often fails to account for the material sources of domination and how social, economic and political structures are produced and transformed (Bina Agarwal, "The Gender and Environment Debate: Lessons from India," *Feminist Studies* 18, no. 1 [1992]: 119–158). Agarwal calls instead for a framework that positions women's relationship to nature as rooted in material reality, structured by a gendered and class-based organization of production, reproduction, and distribution.

42. Caroyln Sachs, *Women Working in the Environment* (Washington, DC: Taylor & Francis, 1997).

43. Bhavnani, Foran, and Kurian, *Feminist Futures*, 14.

44. Fred D'Aguiar, "Wilson Harris," *Bomb Magazine*, January 1, 2003, https://bombmagazine.org/articles/wilson-harris/.

45. Rodney, *A History of the Guyanese Working People*, 102.

46. Cordis, "In Guyana, Colonial Regimes Power the New Oil Frontier," 274.

47. Cordis, "In Guyana, Colonial Regimes Power the New Oil Frontier," 268. Cordis offers a detailed breakdown of US interventionism in Guyana.

48. As stated on GlobeSpan24x7, a political talk show that aired on December 16, 2019, under the title, "Environmental Impact from Oil & Gas," www.youtube.com/watch?v=14fl2EKRW3w.

49. GlobeSpan24x7, "Environmental Impact from Oil & Gas."

50. Description provided by Jan Mangal himself at http://linkedin.com/in/jankmangal/.

51. Jan Mangal, "National Oil Company for Guyana Would Be a Disaster," *Stabroek News* (Guyana), July 2, 2018.

52. Mangal, "National Oil Company for Guyana Would Be a Disaster."

53. "A Fair Deal for Guyana / A Fair Deal for the Planet," https://fairdealforguyana.org/support-the-campaign/.

54. "A Fair Deal for Guyana / A Fair Deal for the Planet."

55. Wilson Harris, "The Landscape of Dreams," in *Wilson Harris: The Uncompromising Imagination*, ed. Hena Maes-Jelinek (Coventry: Dangaroo, 1991), 33.

CHAPTER 4. THE GREAT WALL OF MALÉ

1. *Wave of Change Maldives*, March 2005, video, 03:23 to 03:53, https://www.youtube.com/watch?v=GYT-XSnhw_g.

2. Quoted in "Doorways to the Sea," https://hani-amir.com/blog/2018/10/11/doorways-to-the-sea.

3. Quoted in "Doorways to the Sea."

4. F. R. Cameron, "Saving the 'Disappearing Islands': Climate Change Governance, Pacific Island States and Cosmopolitan Dispositions," *Continuum* 25, no. 6 (2011): 873–86.

5. Uneven development is a geographic and political concept that emerges from a long tradition of Marxist thought. As Niel Smith argues, uneven development is central to the logic of capitalism, and while it is often assumed to be a universal phenomenon, it plays out in place-specific ways (Neil Smith, *Uneven Development: Nature, Capital, and the Production of Space*, 3rd ed. [Athens: University of Georgia Press, 2008]).

6. See Malm, "Sea Wall Politics."

7. Paprocki, "All That Is Solid Melts into the Bay."

8. The Maldives was a British protectorate from 1887 to 1965. The only real presence of the British Empire in the Maldives occurred during World War II, when the British constructed temporary air strips in Addu Atoll, which were abandoned in the 1950s (Uma Kothari and Rorden Wilkinson, "Colonial Imaginaries and Postcolonial Transformations: Exiles, Bases, Beaches," *Third World Quarterly* 31, no. 8 [2010]: 1395–412).

9. Amin Didi, born in 1910, was heir to the sultanate and a descendant of the sixth royal dynasty to rule the country. Amin Didi rejected the throne in a symbolic speech to Parliament, breaking an 850-year cycle of successive rulers. He established the Maldives' first political party in 1950, declaring education for women one of his main goals. On January 1, 1953, after a referendum declared Maldives a republic, the people elected him president. When Amin Didi died on January 19, 1954, the country was returned to his rivals.

10. Crowe, "The Maldive Islands."

11. Ali Shareef, "A Tribute to the Late Ameed Didi," *Addu Diary* [Blog], August 2010, http://addudiary.blogspot.com/2010/08/tribute-to-late-ameen-didi.html.

12. Clarence Mahoney, *People of the Maldive Islands* (Himayatnagar, Hyderabad: Orient Blackswan, 2013), xi.

13. So, *Social Change and Development*, 53.

14. "Yellow Fever: The Maldives Goes to the Polls," *Economist*, September 6, 2013, https://www.economist.com/asia/2013/09/06/yellow-fever.

15. Benjamin K. Sovacool, "Expert Views of Climate Change Adaptation in the Maldives," *Climatic Change* 114, no. 2 (2012): 298–99.

16. Tourist resorts undergo extraordinary environmental alterations to commodify the desire for pristine nature. During my 2013 visit to the Maldives, I toured Robinson Club Resort, a five-star franchise that during the winter holidays charges as much as $4,000 per night. At the time of my visit, the resort was shut down for repairs. The new German owners were in the process of tearing down the lagoon bungalows in order to build new ones. I bore witness to the architectural skeleton of a resort island and watched as tractors dredged sand from the lagoon to fill bags for the island's artificial foundation.

17. Brett Milligan, "Tetrapods, Entropy and Excess," *Free Association Design*, September 18, 2010, https://freeassociationdesign.wordpress.com/2010/09/18/tetrapods-entropy-and-excess/.

18. Hamm, "History of Coastal Engineering in France," 160.

19. CIRIA, CUR, and CETMEF, *The Rock Manual*, 249.

20. This is now surpassed many times over by larger and more ambitious projects funded by other nations with vested interests in the strategic location of the Maldives.

21. Kiyoshi Horikawa, "History of Coastal Engineering in Japan," in Kraus, *History and Heritage of Coastal Engineering*, 336–74.

22. One notable exception occurred in the aftermath of the damage triggered by the 1854 Ansei-Tokai earthquake when a coastal dike was built with private money (Nobuo Shuto and Koji Fujima, "A Short History of Tsunami Research in Japan," *Proceedings of the Japan Academy, Series B Physical and Biological Sciences* 85, no. 8 [2009]: 267–75).

23. Stephen Hesse, "Loving and Loathing Japan's Concrete Coasts, Where Tetrapods Reign," *Asia-Pacific Journal* 5, no. 7 (2007): 1–9.

24. Steve Levenstein, "Plush Tetrapods Turn Your Couch into a Coastline," http://inventorspot.com/articles/plush_tetrapods_turn_your_couch_coastline_24658, accessed April 8, 2023.

25. Sarah Baird, "Covering Coasts with Concrete: Japan Looks to Tetrapods to Battle Elements," Ars Technica, November 25, 2016, https://arstechnica.com/science/2016/11/covering-coasts-with-concrete-japan-looks-to-tetrapods-to-battle-elements/.

26. Shuhei Kimura, "When a Seawall Is Visible: Infrastructure and Obstruction in Post-Tsunami Reconstruction in Japan," *Science as Culture* 25, no. 1 (2016): 23–43.

27. Benjamin K. Sovacool, "Hard and Soft Paths for Climate Change Adaptation," *Climate Policy* 11, no. 4 (2011): 1177–83.

28. Jackson and della Dora, "'Dreams So Big Only the Sea Can Hold Them,'" 2096.

29. As Uma Kothari and Alex Arnall observe, the very sand that is dredged from the seabed is engaged in an elemental placemaking process across multiple temporalities and desires ("Shifting Sands: The Rhythms and Temporalities of Island Sandscapes," *Geoforum* 108 [2020]: 305–14).

30. Alexander K. Naylor, "Island Morphology, Reef Resources, and Development Paths in the Maldives," *Progress in Physical Geography: Earth and Environment* 39, no. 6 (2015): 740.

31. Bisaro et al., "Leveraging Public Adaptation Finance through Urban Land Reclamation," 685.

32. For example, the website Private Islands Online lists islands in the Maldives starting at US$3 million, https://www.privateislandsonline.com/.

33. Jamaica Kincaid, *A Small Place* (New York: Farrar, Straus and Giroux, 1988).

34. *Wave of Change*.

35. Jon Hamilton, "Maldives Builds Barriers to Global Warming," *Morning Edition*, NPR, 2008, https://www.npr.org/templates/story/story.php?storyId=18425626."

36. Naylor, "Island Morphology," 740.

37. Sovacool, "Hard and Soft Paths for Climate Change Adaptation."

38. Naylor, "Island Morphology," 740–41.

39. Sovacool, "Hard and Soft Paths for Climate Change Adaptation," 1179.

40. As a model, economies of scale emerged in the twentieth century with the rise of corporations and came into practice with Fordism.

41. Naylor, "Island Morphology," 745.

42. The infrastructure to bring telecommunications to customers in rural areas is part of a larger challenge. Small and isolated communities are often excluded by major internet service providers because these companies don't stand to profit.

43. Maumoon Abdul Gayoom, "Inaugural Address," Small States Conference on Sea Level Rise, Malé, Republic of Maldives, November 16 1989, 7.

44. Sharon Moshavi, "Tourists Flock to a Sinking Paradise," *Bloomberg Businessweek*, September 17, 1995, http://www.businessweek.com/stories/1995-09-17/tourists-flock-to-a-sinking-paradise-dot-dot-dot-intl-edition.

45. Franny Armstrong, dir., *Drowned Out* (London, 2002). Dams were the most iconic water-based infrastructure projects of the development era, diverting rivers while generating hydroelectric energy. In India, mega dams were regarded as a renewable source of energy that would lift the country out of poverty while bringing water to drought-stricken regions. However, as the film shows, the real function of the dam was to act as a conduit for industrial water needs.

46. Idowu Ajibade, "Planned Retreat in Global South Megacities: Disentangling Policy, Practice, and Environmental Justice," *Climatic Change* 157, no. 2 (2019): 299–317.

47. Isabelle Anguelovski et al., "Equity Impacts of Urban Land Use Planning for Climate Adaptation: Critical Perspectives from the Global North and South," *Journal of Planning Education and Research* 36, no. 3 (2016): 339.

48. "Maldives Tsunami Impact and Recovery: Joint Needs Assessment," Asian Development Bank, the United Nations, and the World Bank (2005), 14.

49. "Sea Wall 'Saves Maldives Capital,'" *BBC News*, January 10, 2005, http://news.bbc.co.uk/2/hi/south_asia/4161491.stm.

50. "Sea Wall 'Saves Maldives Capital.'"

51. Naylor, "Island Morphology."

52. Naylor, "Island Morphology."

53. The safety rationale of Gayoom's plan was met with criticism after people witnessed the differential treatment that his government gave to resorts, which were equally affected but exempt from relocation (Klein, *The Shock Doctrine*, 401).

54. Mimi Sheller, "The Islanding Effect: Post-Disaster Mobility Systems and Humanitarian Logistics in Haiti," *Cultural Geographies* 20, no. 2 (2013): 185–204.

55. Sheller, "The Islanding Effect," 187.

56. Naylor, "Island Morphology."

57. Naylor, "Island Morphology," 741.

58. UN Human Rights Council, *Report of the Special Rapporteur on the Human Rights of Internally Displaced Persons, Addendum: Mission to Maldives,*

January 30, 2012, 12, https://www.refworld.org/docid/4f3935d32.html, accessed April 8, 2023.

59. Klein, *The Shock Doctrine*, 386. Not far from the Maldives, the same tsunami hit the coast of Sri Lanka, devastating the country and claiming the lives of 250,000 people, leaving many more without a home. Prior to the tsunami, the Sri Lankan government's plan to privatize Sri Lanka was met with resistance on the street and at the polls. After the tsunami, "building back better" enabled the government to enact its unpopular agenda.

60. Klein, *The Shock Doctrine*, 389. Ironically, the devastating images of post-tsunami debris and its victims helped generate the funds that would further displace them. This speaks to the danger of the climate refugee image, where funds for loss and damages can be used by governments to displace people or to support widely unpopular development schemes. The power of the refugee narrative to generate sympathy and dollars also carries the potential to deepen existing inequities.

61. Uma Kothari, "Political Discourses of Climate Change and Migration: Resettlement Policies in the Maldives," *Geographical Journal* 180, no. 2 (2014): 130.

62. Carol Farbotko, "Wishful Sinking: Disappearing Islands, Climate Refugees and Cosmopolitan Experimentation," *Asia Pacific Viewpoint* 51, no. 1 (2010): 47–60.

63. Alex Arnall and Uma Kothari, "Challenging Climate Change and Migration Discourse: Different Understandings of Timescale and Temporality in the Maldives," *Global Environmental Change* 31 (2015): 202.

64. Nils-Axel Mörner, Michael Tooley, and Göran Possnert, "New perspectives for the future of the Maldives," *Global and Planetary Change* 40, no. 1 (2004): 177–82.

65. Nils-Axel Mörner, "Claim that Sea Level Is Rising Is a Total Fraud," *Executive Intelligence Review* 34, no. 25 (2007): 35.

66. In it, he wrote, "The people of the Maldives had no problems surviving in the seventeenth century, which was 50cm higher than now. . . . This bodes well for the prospects of surviving the next change. . . . So why the scare-mongering? Could it be because there is money involved?" (Nils-Axel Mörner, "Why the Maldives Aren't Sinking," *Spectator*, December 2, 2009, www.spectator.co.uk/article/why-the-maldives-aren-t-sinking). This was not the first time Mörner had made his views known to the public, in what was clearly becoming an attempt to undermine climate science.

67. Paul Kench, Scott Nichol, and Roger McLean, "Comment on 'New Perspectives for the Future of the Maldives' by Mörner, N. A., et al.," *Global and Planetary Change* 47 (2005): 68.

68. P. S. Kench et al., "Holocene Reef Growth in the Maldives: Evidence of a Mid-Holocene Sea-Level Highstand in the Central Indian Ocean," *Geology* 37, no. 5 (2009): 455–58.

69. Sophie Webber, "Performative Vulnerability: Climate Change Adaptation Policies and Financing in Kiribati," *Environment and Planning A: Economy and Space* 45, no. 11 (2013): 2717–33.

70. Jane McAdam, "'Disappearing States,' Statelessness and the Boundaries of International Law," *UNSW Law Research Paper*, no. 2010–2 (2010): 4, https://ssrn.com/abstract=1539766.

CHAPTER 5. CONTESTED FUTURES: THE HOPE OF A LIVING SEAWALL

1. Formerly under the name Save the Beach Villingili.

2. Kasia Paprocki, "The Climate Change of Your Desires: Climate Migration and Imaginaries of Urban and Rural Climate Futures," *Environment and Planning D: Society and Space* 38, no. 2 (2019): 248.

3. Klein, *The Battle for Paradise*.

4. Ilan Kelman et al., "Does Climate Change Influence People's Migration Decisions in Maldives?," *Climatic Change* 153, no. 1–2 (2019): 293.

5. After polling just 25 percent of the vote in the first round of the elections to Gayoom's 40 percent, Nasheed united the opposition parties in the final round to take 54 percent of the vote and won the presidency.

6. "Aneh Dhivehi Raajje: The Strategic Action Plan. National Framework for Development 2009–2013," Government of the Maldives.

7. One of Nasheed's other priorities was to decentralize tourism. Since 2009, new policies have allowed for the expansion of tourism to populated islands under the assumption that shared access would democratize tourism, allowing for benefits to spread to the Maldivian people, who were previously denied access to the exclusive benefits of tourism in the country. However, despite Nasheed's good intentions, this policy shift has not been entirely fair or just. Kothari and Arnall describe the contradictions of touristic nature as it is being negotiated at the local level between islanders and guesthouse owners who compete over the landscape and its limited beaches and public infrastructure. Interestingly, they found that some islanders expressed resistance to this new model by disrupting and littering on the newly claimed areas cultivated by guesthouse owners (Uma Kothari and Alex Arnall, "Contestation over an Island Imaginary Landscape: The Management and Maintenance of Touristic Nature," *Environment and Planning A: Economy and Space* 49, no. 5 [2017]: 980–98).

8. "Aneh Dhivehi Raajje," 18–19.

9. For more on the defining characteristics of mobility justice, see Mimi Sheller, "Theorising Mobility Justice," *Tempo Social* 30 (2018): 17–34.

10. Broadcasted on December 9, 2011, on *Democracy Now!* under the title, "Frustrated by Inaction, Youth Delegates Occupy COP 17 Plenary in Durban,"

https://www.democracynow.org/2011/12/9/frustrated_by_inaction_youth_delegates_occupy.

11. There is debate about whether Nasheed's development program was progressive enough. Kothari argues that Nasheed rebranded Gayoom's population consolidation agenda, noting that an interisland transit system would similarly encourage islanders to migrate (Kothari, "Political Discourses of Climate Change and Migration"). I share Kothari's critical stance and acknowledge that both plans share common national and international pressures to produce economies of scale, but my interpretation of Nasheed's development program differs. I view Nasheed's plan as oppositional to Gayoom's due to its emphasis on democratic governance, mobility justice, and human rights.

12. An undergraduate thesis by Christopher Gordon documents the strategic use of moral authority in the Maldives to gain geopolitical leverage in the international UN climate change negotiations. See Christopher Gordon, "Planning the Green State on a Sinking Ship: Sea Level Rise Politics in the Maldives" (Honors thesis, Ohio State University, 2010).

13. Benjamin K. Sovacool, "Hard and Soft Paths for Climate Change Adaptation," *Climate Policy* 11, no. 4 (2011): 1180–81.

14. Sovacool, "Expert Views of Climate Change Adaptation in the Maldives," 296.

15. Benjamin K. Sovacool, "Perceptions of Climate Change Risks and Resilient Island Planning in the Maldives," *Mitigation and Adaptation Strategies for Global Change* 17, no. 7 (2012): 742–43.

16. Sovacool, "Expert Views of Climate Change Adaptation in the Maldives," 298.

17. Adam Grydehøj and Ilan Kelman, "The Eco-Island Trap: Climate Change Mitigation and Conspicuous Sustainability," *Area* 49, no. 1 (2017): 106–13.

18. Quoted in Sovacool, "Perceptions of Climate Change Risks," 744.

19. Quoted in Sovacool, "Perceptions of Climate Change Risks," 744.

20. Sovacool, "Expert Views of Climate Change Adaptation in the Maldives," 299.

21. In 2018, after four long years, the MDP found its way back into power with Ibrahim Mohamed Solih as its presidential candidate. The former administration went to great lengths to ensure that Nasheed would not run in that election, arresting him in 2016 and sentencing him to thirteen years in prison, a sentence that was overturned in 2018. While this victory was widely celebrated, trouble was brewing. In May 2021, Nasheed was seriously injured in an assassination attempt outside his home in Malé.

22. The housing minister, Mohammed Muizzu, spoke at a press conference and said, "Due to the current economic situation of the country, 200 residential islands will not be sustainable and that development will only occur if the population of these islands are consolidated" (Azuhaar Abdul Azeez, "Bill on

Population Consolidation Drafted: Housing Minister," *Haveeru Online*, August 6, 2014). *Haveeru Online* was shut down in 2016 by pro-Gayoom supporters under the influence of President Abdulla Yamin.

23. For a fuller context of the events of Nasheed's short-lived victory, see Summer Gray and John Foran, "Climate Injustice: The Real History of the Maldives," *Berkeley Journal of Sociology* 59 (2015): 14–25.

24. Daniel Bosley, "Maldives National Oil Company Seeks Assistance with Oil Exploration," *Minivan News* (Malé, Maldives), February 27, 2014, https://minivannewsarchive.com/environment/maldives-national-oil-company-seeks-assistance-with-oil-exploration-78626.

25. Summer Gray, "Gone Before the Wave" (2015), https://youtube/JpwDVYG3-P8.

26. Neil Merrett, "Maldives Committed to Carbon Neutral Aims Despite Political Uncertainty," *Minivan News* (Malé, Maldives), October 15, 2012, https://minivannewsarchive.com/society/maldives-committed-to-carbon-neutral-aims-despite-political-uncertainty-45440.

27. The Scaling-Up Renewable Energy Program was produced under President Mohamed Nasheed and driven by Nasheed's energy adviser, Mike Mason, a UK national and expert on renewable energy, carbon finance, and carbon offsetting. Due to the unraveling political situation in the country, the plan was suspended. See J. J. Robinson, "Who Turned Out the Light: Maldives' Solar Ambitions Plunged into Darkness," *Minivan News* (Malé, Maldives), June 24, 2012, https://minivannewsarchive.com/politics/who-turned-out-the-light-maldives%E2%80%99-solar-ambitions-plunged-into-darkness-39667.

28. Azra Naseem and Mohamed Mushfique, "The Long Road from Islam to Islamism: A Short History," *Dhivehi Sitee*, May 30, 2014, http://www.dhivehisitee.com/religion/islamism-maldives/.

29. In the Maldives, traditional boat-building relies on oral tradition; there are no blueprints or written plans. One of the traditions that is now prohibited is the celebration that traditionally followed the completion of a boat. Other lost traditions include the celebration of birthdays and death days as well as a host of cultural observances, including *vajidhuvun*, *dhafi negun*, *maaloodhu kiyun*, and *salavaaiy kiyevun*.

30. Posted to the Facebook page of Addu Cultural Center on February 6, 2013, https://www.facebook.com/kuduranaa/.

31. Street art is another place where political dissent is expressed. In the aftermath of the coup, there was an explosion of street art and political messages graffitied onto the seawall surrounding the island of Malé.

32. "Different Strokes: Maldives Power Transfer Expressed through Art," *Minivan News*, August 28, 2012, https://minivannewsarchive.com/politics/different-strokes-maldives-power-transfer-expressed-through-art-42908.

33. "Different Strokes."

34. Ahmed Rilwan, "15 Journalists Receive Death Threats over Gang Reporting," *Minivan News* (Malé, Maldives), August 4, 2014, https://minivannewsarchive.com/crime-2/15-journalists-receive-death-threats-over-gang-reporting-89404.

35. It is also worth asking what counts as a living seawall. Does it have to be designed in a lab? Or can it be unintentional, as in the case of Guyana's riprap structures that happen to attract sea life, in addition to garbage and rodents?

36. Islands are dynamic ecosystems with complex geological characteristics. See Kench et al., "Holocene Reef Growth in the Maldives."

37. Bluepeace Maldives, "Towards an Artificial Island Paradise on Earth," May 24, 2008, http://bluepeacemaldives.org/blog/biodiversity/towards-an-artificial-paradise-on-earth.

38. See François Pyrard, *The Voyage of François Pyrard of Laval to the East Indies, the Maldives, the Moluccas and Brazil*, vol. 3, ed. Albert Gray and H. C. P. Bell (Cambridge: Cambridge University Press, 2010).

39. Naylor, "Island Morphology," 735.

40. Naylor, "Island Morphology," 736.

41. Naylor, "Island Morphology," 737.

42. Naylor, "Island Morphology," 738.

43. Morita, "Infrastructuring Amphibious Space," 135.

44. Morita, "Infrastructuring Amphibious Space," 135.

45. Morita, "Infrastructuring Amphibious Space," 108.

46. There is an irony that is hard to ignore in the use of concrete to construct artificial reefs, as the making of cement contributes significant concentrations of CO_2 to the atmosphere. Given the artificial nature of living seawalls, advocates of artificial reefs remain caught in the awkward politics of eco-minimalism, as designers are faced with mulling over ways to minimize their carbon footprint so as to offset the externalities of their creations.

47. While coral planting and 3D printing have generated excitement among some members of the tourism industry, it is often said that tourists can't tell the difference between a vibrant, living reef and one that has been bleached to death.

48. Cronon, "The Trouble with Wilderness."

49. Kothari and Arnall, "Contestation over an Island Imaginary Landscape," 985.

50. Kothari and Wilkinson, "Colonial Imaginaries and Postcolonial Transformations."

51. The libertarian Seasteading movement aims to build communities on floating cities in autonomous international waters. For a critical analysis of the Seasteading vision, see Philip E. Steinberg, Elizabeth Nyman, and Mauro J. Caraccioli, "Atlas Swam: Freedom, Capital, and Floating Sovereignties in the Seasteading Vision," *Antipode* 44, no. 4 (2012).

52. Ahdha Moosa, Khoa Do, and Emil Jonescu, "Design Response to Rising Sea Levels in the Maldives: A Study Into Aquatic Architecture," *Frontiers of Architectural Research* 9, no. 3 (2020): 627, 632.

53. For a more detailed account of green gentrification, see Isabelle Anguelovski and James J. T. Connolly, *The Green City and Social Justice: 21 Tales from North America and Europe* (New York: Routledge, 2022).

54. Patrick D. Nunn, "Understanding and Adapting to Sea-Level Rise," in *Global Environmental Issues*, ed. Frances Harris (Oxford: Wiley-Blackwell, 2012), 87–104.

55. Jon Shenk, *The Island President* (First Run Features, March 30, 2012), DVD. Jon Shenk follows the story of Nasheed in some detail, including the torture and repression that characterized Gayoom's thirty-year rule prior to Nasheed's 2008 rise to power. Shenk would eventually go on to direct *An Inconvenient Sequel*, following Al Gore in a similar fashion.

56. For more on history and popular culture in the Maldives, see Xavier Romero-Frías, *The Maldive Islanders: A Study of the Popular Culture of an Ancient Ocean Kingdom* (Barcelona: Nova Ethnographia Indica, 2003).

CONCLUSION: THE DILEMMA OF PLACEKEEPING

1. Kathy Jetñil-Kijiner, "Dear Matafele Peinem" (2014), https://kathyjetnilkijiner.com.

2. Winner, "Do Artifacts Have Politics?"

3. Susan Leigh Star, "The Ethnography of Infrastructure," *American Behavioral Scientist* 43, no. 3 (1999): 377–91.

4. Pilkey and Young, *The Rising Sea*.

5. Pilkey and Young, *The Rising Sea*, 215.

6. Jennifer L. Rice, Joshua Long, and Anthony Levenda, "Against Climate Apartheid: Confronting the Persistent Legacies of Expendability for Climate Justice," *Environment and Planning E: Nature and Space* 5, no. 2 (2022): 625–45.

7. See Matthew Thomas Clement, "'Let Them Build Sea Walls': Ecological Crisis, Economic Crisis and the Political Economic Opportunity Structure," *Critical Sociology* 37, no. 4 (2011): 447–63.

8. Mona Lalwani, "A Danish Company Is Building a $335 Million Seawall around New York," *The Verge*, October 1, 2014, https://www.theverge.com/2014/10/1/6874925/can-a-massive-seawall-save-new-york-from-flooding.

9. Vanessa Quirk, "The BIG U: BIG's New York City Vision for 'Rebuild by Design,'" *ArchDaily*, April 4, 2014, https://www.archdaily.com/493406/the-big-u-big-s-new-york-city-vision-for-rebuild-by-design.

10. E. Melanie DuPuis and Miriam Greenberg, "The Right to the Resilient City: Progressive Politics and the Green Growth Machine in New York City," *Journal of Environmental Studies and Sciences* 9, no. 3 (2019): 352–63.

11. Alan Feuer, "Protecting the City, Before Next Time," *New York Times*, November 3, 2012, https://www.nytimes.com/2012/11/04/nyregion/protecting-new-york-city-before-next-time.html.

12. Sylvia Poggioli, "MOSE Project Aims to Part Venice Floods," NPR, January 7, 2008, https://www.npr.org/2008/01/07/17855145/mose-project-aims-to-part-venice-floods.

13. Climate change also translates into more frequent storm surges and the deterioration of lagoon sediments, which threatens what remains of the natural barriers formed by salt marshes and mudflats. Furthermore, MOSE is not designed to account for long-term sea change.

14. R. W. Kates et al., "Reconstruction of New Orleans after Hurricane Katrina: A Research Perspective," *Proceedings of the National Academy of Sciences* 103, no. 40 (2006): 14653–60.

15. Benjamin Sims, "Things Fall Apart: Disaster, Infrastructure, and Risk," *Social Studies of Science* 37, no. 1 (2007): 94. For additional insight on the causes and consequences of infrastructure repair, see Christopher R. Henke and Benjamin Sims, *Repairing Infrastructures: The Maintenance of Materiality and Power* (Cambridge, MA: MIT Press, 2020).

16. It is widely known that unregulated tourism and the expansion of port activity lies at the heart of the injustice faced by Venetians who are trying to safeguard the city for future inhabitants. This is in part due to the recent acquiescence of the government to the demands of the cruise industry, which has been granted the privilege of nearly unrestricted access to the lagoon. When Venice was granted the status of a UNESCO World Heritage Site in 1987, the Venetian government became responsible for developing a management plan. The plan to safeguard the city, presented to the public only in 2012, has systematically ignored the "vicious circle" of tourism (Antonio Paolo Russo, "The 'Vicious Circle' of Tourism Development in Heritage Cities," *Annals of Tourism Research* 29, no. 1 [2002]: 165–82). In 2009, weeks before the climate change negotiations were set to take place in Copenhagen (COP15), Venetians held a mock funeral to highlight their dwindling population and eventual disappearance.

17. The Orsoni government has been accused of inflating the price of MOSE by 30 percent and handing out corrupt contracts. In 2013, Giorgio Orsoni, mayor of Venice, and thirty-four officials were arrested on charges of embezzling millions of dollars in public funds from MOSE.

18. Lidia Panzeri and Anna Somers Cocks, "Huge Financial Scandal in Venice," *Art Newspaper*, July 15, 2013.

19. Clement, "'Let Them Build Seawalls.'" The phrase is adapted from the work of John Bellamy Foster, *Ecology Against Capitalism* (New York: Monthly Review Press, 2002), 65.

20. Indonesia, for example, recently launched a $263 million seawall project, to be constructed with the help of South Korea, builder of the world's largest seawall earlier this decade. In Jakarta, water pumping has sunk the land below sea level, endangering ten million people and justifying the need for a seawall. Once completed, the wall will transform the city's landscape in favor of private development, displacing much of the existing population.

21. Summer Gray et al., "Climate Justice Movements and Sustainable Development Goals," in *Climate Action*, ed. Walter Leal Filho et al. (Cham: Springer International, 2020), 1–10.

22. There are some noteworthy efforts to expand adaptation framings, but these inventions often assume that better policy frameworks can help achieve more just and equitable adaptation without challenging the drivers and underlying conditions of vulnerability. For example, "just adaptation" is a school of thought that concerns the implementation of climate adaptation impacts among socially vulnerable groups. It is better than conventional adaptation frameworks but is limited to the standard legal understandings of justice-related concerns, for example, distributive equity, social recognition, and public participation or procedural justice, therefore overlooking other dimensions of justice such as interactional, spatial, and temporal fairness. See Sonia Graham et al., "Towards Fair Local Outcomes in Adaptation to Sea-Level Rise," *Climatic Change* 130, no. 3 (2015): 411–24.

23. See the work of Andrew W. Kahrl, including *Free the Beaches: The Story of Ned Coll and the Battle for America's Most Exclusive* (New Haven, CT: Yale University Press, 2018); and *The Land Was Ours: How Black Beaches Became White Wealth in the Coastal South* (Chapel Hill: University of North Carolina Press, 2016).

24. For example, in May 1945, Black activists held a "wade-in" at a whites-only beach in Miami to protest Jim Crow–era laws (Gregory W. Bush, *White Sand, Black Beach: Civil Rights, Public Space, and Miami's Virginia Key* [Gainesville: University Press of Florida, 2016]). Beach apartheid in South Africa similarly came under pressure in the 1980s. In 1989, a people's campaign rallied around slogans like "Drown Beach Apartheid" and culminated in protest picnics on segregated beaches and actions to "flood" white beaches with Black bodies (Jayne M. Rogerson, "'Kicking Sand in the Face of Apartheid': Segregated Beaches in South Africa," *Bulletin of Geography* 35, no. 35 [2017]: 93–110). In 2010, when New Zealand granted permission to an outside company to conduct seismic testing and offshore exploration, Māori communities protested by holding beach bonfires and direct-action flotillas (Patricia Widener, "Coastal People Dispute Offshore Oil Exploration: Toward a Study of Embedded Seascapes, Submersible Knowledge, Sacrifice, and Marine Justice," *Environmental Sociology* 4, no. 4 [2018]: 405–18).

25. Hardy, Milligan, and Heynen, "Racial Coastal Formation," 63.

26. Rebecca Elliott, "The Sociology of Climate Change as a Sociology of Loss."

27. Mimi Sheller, *Island Futures: Caribbean Survival in the Anthropocene* (Durham, NC: Duke University Press, 2020).

28. Paprocki, "The Climate Change of Your Desires."

29. Ranganathan and Bratman, "From Urban Resilience to Abolitionist Climate Justice in Washington, DC," 121.

30. Ranganathan and Bratman, "From Urban Resilience to Abolitionist Climate Justice in Washington, DC," 132.

31. In 2008, residents of the small Alaska native village of Kivalina filed a legal claim against ExxonMobil for damaging their homeland and creating a false debate about climate change (Christine Shearer, *Kivalina: A Climate Change Story* [Chicago: Haymarket, 2011]). In Shishmaref, Alaska, frontline Indigenous communities have been engaged in a prolonged battle with the federal government for relocation funds (Elizabeth K. Marino, *Fierce Climate, Sacred Ground: An Ethnography of Climate Change in Shishmaref Alaska* [Fairbanks: University of Alaska Press, 2015]). Small island states have similarly fought for mitigation and funding for loss and damages (Sam Adelman, "Climate Justice, Loss and Damage and Compensation for Small Island Developing States," *Journal of Human Rights and the Environment* 7, no. 1 [2016]: 32–53). In 2019, coastal justice activists succeeded in integrating the principles of environmental justice and social equity into the language of coastal policy in California.

32. Ajibade, "Planned Retreat in Global South Megacities," 312.

33. If climate justice leaves room for place and attention to historical and intersectional processes on the ground, this space narrows when the focus shifts to the international scale, where the discourse leans heavily toward the fairness dimension of distributive justice; here little attention has been given to the local impacts of adaptation practices (Graham et al., "Towards Fair Local Outcomes in Adaptation to Sea-Level Rise," 412). This is not to argue that climate justice is not a critical and necessary intervention in the business-as-usual politics of the moment, but there is a need to explore the different usages of the term and to what extent climate justice can fully account for the place-specific harms that exist now and will exist in the future, particularly for those who live on the coast.

34. "Managed retreat" refers to the process of relocating people and infrastructure inland, away from coastal hazards. In addition to the protect/retreat dichotomy, many assume that managed retreat is not a form of adaptation. As Liz Koslov and others argue, relocation should be considered adaptation, not a failure to adapt (Liz Koslov, "The Case for Retreat," *Public Culture* 28 no. 2 [2016]: 359–87).

35. For example, FEMA and HUD use cost-benefit analysis. For a comprehensive list of funding agencies and rules, see A. R. Siders, "Social Justice Implications of US Managed Retreat Buyout Programs," *Climatic Change* 152, no. 2 (2019): 239–57.

36. Anamaria Bukvic, "Leaving the Shore: What Drives Decision-Making after Coastal Disasters," Natural Hazards Center, November 11, 2020, https://hazards.colorado.edu/news/research-counts/leaving-the-shore-what-drives-decision-making-after-coastal-disasters.

37. Siders, "Social Justice Implications of US Managed Retreat Buyout Programs," 239.

38. Ajibade, "Planned Retreat in Global South Megacities."

39. Even in the Global North, managed retreat has the tendency to reinforce existing inequalities. Institutionally, managed retreat stems from a contested history of disaster retreat, a process that occurs only *after* a major disaster occurs. Studies have shown that the immediacy of disaster can shift priorities from the established criteria of helping low- and moderate-income households in favor of wealthier, high-density communities (Ajibade, "Planned Retreat in Global South Megacities"). While the priorities of disaster risk reduction and climate change adaptation share many priorities, they operate in different temporalities and scales, defining hazards differently.

40. For example, see Anguelovski et al., "Equity Impacts of Urban Land Use Planning for Climate Adaptation," 339.

41. Kril and Roopnaraine, *The Seawall: Tales of the Guyana Coast.*

42. Naomi Klein, *This Changes Everything: Capitalism vs. the Climate* (New York: Simon & Schuster, 2015), 351.

43. Malm, "Sea Wall Politics," 827.

44. DuPuis and Greenberg, "The Right to the Resilient City."

45. Jetñil-Kijiner and Niviâna, "Rise: From One Island to Another."

46. Joanna Macy, *Pass it On: Five Stories That Can Change the World* (Berkeley, CA: Parallax Press, 2010), 97.

Bibliography

Adelman, Sam. "Climate Justice, Loss and Damage and Compensation for Small Island Developing States." *Journal of Human Rights and the Environment* 7, no. 1 (2016): 32–53.

Adorno, Theodor W., and Max Horkheimer. *Dialectic of Enlightenment*. Translated by Edmund Jephcott. Edited by Gunzelin Schmid Noerr. Stanford, CA: Stanford University Press, 2002. First published 1944 as *Philosophical Fragments* by Social Studies Association (New York).

Agarwal, Bina. "The Gender and Environment Debate: Lessons from India." *Feminist Studies* 18, no. 1 (1992): 119–158.

Agrawal, Arun, and Clark C. Gibson. "Enchantment and Disenchantment: The Role of Community in Natural Resource Conservation." *World Development* 27, no. 4 (1999): 629–49.

Ajibade, Idowu. "Planned Retreat in Global South Megacities: Disentangling Policy, Practice, and Environmental Justice." *Climatic Change* 157, no. 2 (2019): 299–317.

Alexander, D. E. "Resilience and Disaster Risk Reduction: An Etymological Journey." *Natural Hazards and Earth System Sciences* 13, no. 11 (2013): 2707–16.

Amin, Ash. "Lively Infrastructure." *Theory, Culture & Society* 31, no. 7–8 (2014): 137–61.

Anand, Nikhil, Akhil Gupta, and Hannah Appel, eds. *The Promise of Infrastructure*. Durham, NC: Duke University Press, 2018.

"Aneh Dhivehi Raajje: The Strategic Action Plan. National Framework for Development 2009–2013." Government of Maldives, 2009. http://planning.gov.mv (accessed November 26, 2014).

Anguelovski, Isabelle, and James J. T. Connolly. *The Green City and Social Justice: 21 Tales from North America and Europe*. New York: Routledge, 2022.

Anguelovski, Isabelle, James J. T. Connolly, Hamil Pearsall, Galia Shokry, Melissa Checker, Juliana Maantay, Kenneth Gould, et al. "Why Green 'Climate Gentrification' Threatens Poor and Vulnerable Populations." *Proceedings of the National Academy of Sciences of the United States of America* 116, no. 52 (2019): 26139–43.

Anguelovski, Isabelle, Linda Shi, Eric Chu, Daniel Gallagher, Kian Goh, Zachary Lamb, Kara Reeve, and Hannah Teicher. "Equity Impacts of Urban Land Use Planning for Climate Adaptation: Critical Perspectives from the Global North and South." *Journal of Planning Education and Research* 36, no. 3 (2016): 333–48.

Anthony, Edward J., and Nicolas Gratiot. "Coastal Engineering and Large-Scale Mangrove Destruction in Guyana, South America: Averting an Environmental Catastrophe in the Making." *Ecological Engineering* 47 (2012): 268–73.

Antunes, Catia. "Early Modern Ports, 1500–1750." *European History Online (EGO)* (2010), http://worldcat.org. https://nbn-resolving.org/urn:nbn:de:0159-2010102547.

Armstrong, Franny, dir. *Drowned Out*. London, 2002.

Arnall, Alex, and Uma Kothari. "Challenging Climate Change and Migration Discourse: Different Understandings of Timescale and Temporality in the Maldives." *Global Environmental Change* 31 (2015): 199–206.

Asian Development Bank, the United Nations, and the World Bank. *Maldives Tsunami Impact and Recovery: Joint Needs Assessment*. 2005.

Azeez, Azuhaar Abdul. "Bill on Population Consolidation Drafted: Housing Minister." *Haveeru Online*, August 6, 2014.

Baichwal, Jennifer, and Edward Burtynsky, dir. *Watermark*. New York, 2012.

Baird, Sarah. "Covering Coasts with Concrete: Japan Looks to Tetrapods to Battle Elements." Ars Technica, November 25, 2016. https://arstechnica.com/science/2016/11/covering-coasts-with-concrete-japan-looks-to-tetrapods-to-battle-elements/.

Bankoff, Greg. "Remaking the World in Our Own Image: Vulnerability, Resilience and Adaptation as Historical Discourses." *Disasters* 43, no. 2 (2019): 221–39.

Bayley, Geo D. *Handbook of British Guiana, 1909*. London: Forgotten Books, 2018.

Belsky, Jill M. "Gender, Community, and the Politics of Community-Based Rural Ecotourism in Belize." In *Contested Nature: Promoting International Biodiversity Conservation with Social Justice in the Twenty-First Century*, edited by Steven R. Brechin, Peter R. Wilshusen, Crystal L. Fortwangler and Patrick C. West, 89–102. New York: State University of New York Press, 2003.

Bennett, Jane. "The Agency of Assemblages and the North American Blackout." *Public Culture* 17 (October 1, 2005): 445–66.

———. *Vibrant Matter: A Political Ecology of Things*. Durham, NC: Duke University Press, 2010.

Bennett, Tony, Lawrence Grossberg, Meaghan Morris, and Raymond Williams, eds. *New Keywords: A Revised Vocabulary of Culture and Society*. Malden, MA: Blackwell, 2005.

Bhattacharya, Rahul. *The Sly Company of People Who Care*. London: Pan Macmillan, 2011.

Bhavnani, Kum-Kum, Peter Chua, and Dana Collins. "Critical Approaches to Qualitative Research." In *The Oxford Handbook of Qualitative Research*, edited by Patricia Leavy, 165–78. New York: Oxford University Press, 2014.

Bhavnani, Kum-Kum, John Foran, and Priya Kurian, eds. *Feminist Futures: Re-Imagining Women, Culture and Development*. London: Zed Books, 2003.

Bisaro, Alexander, Mark de Bel, Jochen Hinkel, Sien Kok, and Laurens M. Bouwer. "Leveraging Public Adaptation Finance through Urban Land Reclamation: Cases from Germany, the Netherlands and the Maldives." *Climatic Change* 160, no. 4 (2020): 671–89.

Bluepeace Maldives. "Toward an Artificial Paradise on Earth." 2008. http://bluepeacemaldives.org/blog/biodiversity/towards-an-artificial-paradise-on-earth.

Brechin, Steven R., Peter R. Wilshusen, Crystal L. Fortwangler, and Patrick C. West, eds. *Contested Nature: Promoting International Biodiversity and Social Justice in the Twenty-First Century*. New York: State University of New York Press, 2003.

Browder, Greg, Suzanne Ozment, Irene Rehberger Bescos, Todd Gartner, and Glenn-Marie Lange. *Integrating Green and Gray: Creating Next Generation Infrastructure*. Washington, DC: World Bank and World Resources Institute, 2019.

Bukvic, Anamaria. "Leaving the Shore: What Drives Decision-Making after Coastal Disasters." Natural Hazards Center, November 11, 2020. https://hazards.colorado.edu/news/research-counts/leaving-the-shore-what-drives-decision-making-after-coastal-disasters.

Burawoy, Michael, Joseph A. Blum, Sheba Goerge, Zsuzsa Gille, and Millie Thayer, eds. *Global Ethnography: Forces, Connections, and Imaginations in a Postmodern World*. Berkeley: University of California Press, 2000.

Bush, Gregory W. *White Sand Black Beach: Civil Rights, Public Space, and Miami's Virginia Key*. Gainesville: University Press of Florida, 2016.

Butler, Octavia. *Parable of the Sower*. New York: Warner, 1993.

Cameron, F. R. "Saving the 'Disappearing Islands': Climate Change Governance, Pacific Island States and Cosmopolitan Dispositions." *Continuum* 25, no. 6 (2011): 873–86.

Carson, Rachel. *The Edge of the Sea*. New York: Houghton Mifflin, 1998. First published 1955.

——. *Silent Spring*. 50th anniversary ed. New York: Houghton Mifflin, 2002.

Case, Edward. "The Dymchurch Wall and Reclamation of Romney Marsh." *The Engineer* 88 (1899): 407–8.

Case, Gerald O. *Coast Erosion Protection Works on the Case System in British Guiana*. New York: Tidal Engineering Corporation, 1920.

Certeau, Michel de. *The Practice of Everday Life*. Translated by Steven Rendall. Berkeley: University of California Press, 1984.

Chandler, David. "Resilience and the End(s) of the Politics of Adaptation." *Resilience* 7, no. 3 (2019): 1–10.

Chase, Ashton. *Guyana—a Nation in Transit: Burnham's Role*. Georgetown, Guyana: Paunik, 1994.

Cheltenham, Richard L., Seenath Jairam, and Jacqueline Samuels-Browne. *Report of the Commission of Inquiry Appointed to Enquire and Report on the Circumstances Surrounding the Death of the Late Dr. Walter Rodney on Thirteenth Day of June, One Thousand Nine Hundred and Eighty at Georgetown*. Parliament of the Co-operative Republic of Guyana, 2016. http://parliament.gov.gy/documents/documents-laid/.

CIRIA, CUR, and CETMEF. *The Rock Manual: The Use of Rock in Hydraulic Engineering*. 2nd ed. London: Construction Industry Research & Information Association, 2007.

Clement, Matthew Thomas. "'Let Them Build Sea Walls': Ecological Crisis, Economic Crisis and the Political Economic Opportunity Structure." *Critical Sociology* 37, no. 4 (2011): 447–63.

Cordis, Shanya. "Forging Relational Difference: Racial Gendered Violence and Dispossession in Guyana." *Small Axe: A Caribbean Journal of Criticism* 23, no. 3 (2019): 18–33.

——. "In Guyana, Colonial Regimes Power the New Oil Frontier." *NACLA Report on the Americas* 53, no. 3 (2021): 268–74.

Coss-Corzo, Alejandro De. "Patchwork: Repair Labor and the Logic of Infrastructure Adaptation in Mexico City." *Environment and Planning D: Society and Space* 39, no. 2 (2021): 237–53.

Cousins, Joshua J. "Justice in Nature-Based Solutions: Research and Pathways." *Ecological Economics* 180 (2021). https://doi.org/10.1016/j.ecolecon.2020.106874.

Cronon, William. "The Trouble with Wilderness: Or, Getting Back to the Wrong Nature." *Environmental History* 1, no. 1 (1996): 7–28.

Crowe, Philip. "The Maldive Islands." In *Diversions of a Diplomat in Ceylon*, edited by Philip K. Crowe, 280–304. New York: D. Van Nostrand Co., 1956.

D'Aguiar, Fred. "Wilson Harris." *Bomb Magazine*, January 1, 2003. http://bombmagazine.org/article/2537/wilson-harris.

Dalton, Henry G. *The History of British Guiana*. Vol. 1. London: Longman, Brown, Green, and Longmans, 1855.

Daly, Vere T. *The Making of Guyana*. London: Macmillan, 1974.

Davies, Archie. "The Coloniality of Infrastructure: Engineering, Landscape and Modernity in Recife." *Environment and Planning D: Society and Space* 39, no. 4 (2021): 740–57.

Davoudi, Simon. "Resilience: A Bridging Concept or a Dead End?" *Planning Theory & Practice* 13, no. 2 (2012): 299–333.

DellaSala, Dominick A. "'Real' vs. 'Fake' Forests: Why Tree Plantations Are Not Forests." In *Encyclopedia of the World's Biomes*, edited by Michael I. Goldstein and Dominick A. DellaSala, 47–55. Oxford: Elsevier, 2020.

Dempsey, Nicola, Harry Smith, and Mel Burton, eds. *Place-Keeping: Open Space Management in Practice*. New York: Routledge, 2014.

Dicken, Peter. *Global Shift: Mapping the Changing Contours of the World Economy*. 5th ed. London: Sage, 2007.

Donzinger, Steven, Laura Garr, and Aaron Marr Page. "Rainforest Chernobyl Revisited: The Clash of Human Rights and Bit Investor Claims: Chevron's Abusive Litigation in Ecuador's Amazon." *Human Rights Brief* 17, no. 2 (2010): 1–8.

Dowie, Mark. *Conservation Refugees: The Hundred-Year Conflict between Global Conservation and Native Peoples*. Cambridge, MA: MIT Press, 2009.

DuPuis, E. Melanie, and Miriam Greenberg. "The Right to the Resilient City: Progressive Politics and the Green Growth Machine in New York City." *Journal of Environmental Studies and Sciences* 9, no. 3 (2019): 352–63.

Elliott, Rebecca. "The Sociology of Climate Change as a Sociology of Loss." *European Journal of Sociology* 59, no. 3 (2018): 301–37.

Environmental Protection Agency and Ministry of Natural Resources and the Environment Georgetown. "Fifth National Report to the Convention on Biological Diversity." September 2014. https://www.cbd.int/doc/world/gy/gy-nr-05-en.pdf.

Escobar, Arturo. "Planning." In *The Development Dictionary: A Guide to Knowledge as Power*, edited by Wolfgang Sachs, 145–60. New York: Zed Books, 2010.

Esteva, Gustavo. "Development." In *The Development Dictionary: A Guide to Knowledge as Power*, edited by Wolfgang Sachs, 1–23. New York: Zed Books, 2010.

European Commission, Directorate-General for International Cooperation and Development. *Using Innovative and Effective Approaches to Deliver Climate Change Support to Developing Countries*. Publications Offfice, 2012. https://data.europa.eu/doi/10.2783/76697.

Fainstein, Susan. "Resilience and Justice." *International Journal of Urban and Regional Research* 39, no. 1 (2015): 157–67.

Farbotko, Carol. "Wishful Sinking: Disappearing Islands, Climate Refugees and Cosmopolitan Experimentation." *Asia Pacific Viewpoint* 51, no. 1 (2010): 47–60.

Foster, John Bellamy. "Capitalism and Ecology: The Nature of the Contradiction." Paper presented at the Socialism Conference, Chicago, 2002.

———. *Ecology Against Capitalism*. New York: Monthly Review Press, 2002.
Fougère, Martin, and Eija Meriläinen. "Exposing Three Dark Sides of Social Innovation through Critical Perspectives on Resilience." *Industry and Innovation* 28, no. 1 (2021): 1–18.
Frank, Andre Gunder. "The Development of Underdevelopment." *Monthly Review* 18, no. 4 (1966): 17–31.
Gayoom, Maumoon Abdul. "Inaugural Address." Presented at the Small States Conference on Sea Level Rise, Malé, Republic of Maldives, November 16, 1989.
Ghosh, Amitav. *The Hungry Tide*. New York: Mariner Books, 2006.
Gidel, Mélanie. "Fragmentation on the Waterfront: Coastal Squatting Settlements and Urban Renewal Projects in the Caribbean." In *Transforming Urban Waterfronts: Fixity and Flow*, edited by Gene Desfor, Jennefer Laidley, Quentin Stevens, and Dirk Schubert, 35–53. New York: Routledge, 2011.
Gieryn, Thomas F. "A Space for Place in Sociology." *Annual Review of Sociology* 26, no. 1 (2000): 463–96.
Glavovic, B.C., R. Dawson, W. Chow, M. Garschagen, M. Haasnoot, C. Singh, and A. Thomas. "Cross-Chapter Paper 2: Cities and Settlements by the Sea." In *Climate Change 2022: Impacts, Adaptation, and Vulnerability. Contribution of Working Group II to the Sixth Assessment Report of the Intergovernmental Panel on Climate Change*, edited by H.-O. Pörtner, D. C. Roberts, E. S. Poloczanska, K. Mintenbeck, M. Tignor, A. Alegría, M. Craig, S. Langsdorf, S. Löschke, V. Möller, A. Okem, and B. Rama, 2163–94. Cambridge: Cambridge University Press, 2022.
Gonzales, Rosa. "Community-Driven Climate Resilience Planning: A Framework, Version 2.0." National Association of Climate Resilience Planners, 2017. https://movementstrategy.org/resources/community-driven-climate-resilience-planning-a-framework/.
Gordon, Christopher. "Planning the Green State on a Sinking Ship: Sea Level Rise Politics in the Maldives." Honors thesis, Ohio State University, 2010.
Gourlay, Michael R. "History of Coastal Engineering in Australia." In *History and Heritage of Coastal Engineering: A Collection of Papers on the History of Coastal Engineering in Countries Hosting the International Coastal Engineering Conference 1950–1996*, edited by Nicholas C. Kraus, 1–88. New York: American Society of Civil Engineers, 1996.
Graham, Sonia, Jon Barnett, Ruth Fincher, Colette Mortreux, and Anna Hurlimann. "Towards Fair Local Outcomes in Adaptation to Sea-Level Rise." *Climatic Change* 130, no. 3 (2015): 411–24.
Gray, Summer, and John Foran. "Climate Injustice: The Real History of the Maldives." *Berkeley Journal of Sociology* 59 (2015): 14–25.
Gray, Summer, Corrie Grosse, Brigid Mark, and Erica Morrell. "Climate Justice Movements and Sustainable Development Goals." In *Climate Action*, edited by Walter Leal Filho, Anabela Marisa Azul, Luciana Brandli, Pinar Gökcin Özuyar, and Tony Wall, 1–10. Cham: Springer International, 2020.

Grove, Kevin, Savannah Cox, and Allain Barnett. "Racializing Resilience: Assemblage, Critique, and Contested Futures in Greater Miami Resilience Planning." *Annals of the American Association of Geographers* 110, no. 5 (2020): 1613–30.

Grydehøj, Adam, and Ilan Kelman. "The Eco-Island Trap: Climate Change Mitigation and Conspicuous Sustainability." *Area* 49, no. 1 (2017): 106–13.

Guha, Ramachandra, and Juan Martinez-Alier. *Varieties of Environmentalism: Essays North and South*. London: Earthscan, 2006.

Hall, Stuart. "Cultural Identity and Diaspora." *Framework: The Journal of Cinema and Media* 36 (1989): 222–37.

Hamilton, Jon. "Maldives Builds Barriers to Global Warming." *Morning Edition*, NPR, 2008. https://www.npr.org/templates/story/story.php?storyId=18425626.

Hamm, Luc. "History of Coastal Engineering in France." In *History and Heritage of Coastal Engineering: A Collection of Papers on the History of Coastal Engineering in Countries Hosting the International Coastal Engineering Conference 1950–1996*, edited by Nicholas C. Kraus, 142–68. New York: American Society of Civil Engineers, 1996.

Hardy, R. Dean, Richard A. Milligan, and Nik Heynen. "Racial Coastal Formation: The Environmental Injustice of Colorblind Adaptation Planning for Sea-Level Rise." *Geoforum* 87 (2017): 62–72.

Harris, Willson. "A Note on the Genesis of the Guyana Quartet." In *The Guyana Quartet*, 7–14. London: Faber and Faber, 1985.

———. "The Landscape of Dreams." In *Wilson Harris: The Uncompromising Imagination*, edited by Hena Maes-Jelinek, 31–38. Coventry: Dangaroo, 1991.

Haushofer, Marlen. *The Wall*. Translated by Shaun Whiteside. Pittsburgh, PA: Cleis, 1990.

Henke, Christopher R., and Benjamin Sims. *Repairing Infrastructures: The Maintenance of Materiality and Power*. Cambridge, MA: MIT Press, 2020.

Hesse, Stephen. "Loving and Loathing Japan's Concrete Coasts, Where Tetrapods Reign." *Asia-Pacific Journal* 5, no. 7 (2007): 1–9.

Horikawa, Kiyoshi. "History of Coastal Engineering in Japan." In *History and Heritage of Coastal Engineering: A Collection of Papers on the History of Coastal Engineering in Countries Hosting the International Coastal Engineering Conference 1950–1996*, edited by Nicholas C. Kraus, 336–74. New York: American Society of Civil Engineers, 1996.

Huvila, Isto. "The Politics of Boundary Objects: Hegemonic Interventions and the Making of a Document." *Journal of the American Society for Information Science and Technology* 62, no. 12 (2011): 2528–39.

IPCC. "Summary for Policymakers." In *Climate Change 2022: Impacts, Adaptation, and Vulnerability. Contribution of Working Group II to the Sixth Assessment Report of the Intergovernmental Panel on Climate Change*, edited by H.-O. Pörtner, D. C. Roberts, E. S. Poloczanska, K. Mintenbeck, M. Tignor, A. Alegría, M. Craig, S. Langsdorf, S. Löschke, V. Möller, A. Okem, and B. Rama, 3–33. Cambridge: Cambridge University Press, 2022.

Ishmael, Odeen. *The Guyana Story: From Earliest Times to Independence.* Bloomington, IN: Xlibris, 2013.

Jackson, Mark, and Veronica della Dora. "'Dreams So Big Only the Sea Can Hold Them': Man-Made Islands as Anxious Spaces, Cultural Icons, and Travelling Visions." *Environment and Planning A: Economy and Space* 41 (2009): 2086–104.

Jagan, Cheddi. "Death to Imperialism." Cheddi Jagan Research Center, 1954. https://jagan.org/CJ%20Articles/Early%20Aricles/early_cj.html.

———. *Fight for Freedom: Waddington Constitution Exposed.* Georgetown, Guyana: C. Jagan, 1952.

Jagan, Janet. "A Piece of Guyana's History, 1953–1955." *Thunder: Theoretical and Analytical Journal of the People's Progressive Party of Guyana* 15, no. 1 (1983). Available at https://jagan.org/Janet%20Jagan/JJ%20Articles/images/7259.pdf.

Jetñil-Kijiner, Kathy. "Dear Matafele Peinem." 2014. https://kathyjetnilkijiner.com.

Jetñil-Kijiner, Kathy, and Aka Niviâna. "Rise: From One Island to Another." 350.org, 2018. https://350.org/rise-from-one-island-to-another.

Joerges, Bernward. "Do Politics Have Artefacts?" *Social Studies of Science* 29, no. 3 (1999): 411–31.

Jordan, Michael. "Dave Martins Guyanese Cultural Icon." *Kaieteur News*, May 9, 2010. www.kaieteurnewsonline.com/2010/05/09/dave-martins-guyanese-cultural-icon/.

Kahrl, Andrew W. *Free the Beaches: The Story of Ned Coll and the Battle for America's Most Exclusive Shoreline.* New Haven, CT: Yale University Press, 2018.

———. *The Land Was Ours: How Black Beaches Became White Wealth in the Coastal South.* Chapel Hill: University of North Carolina Press, 2016.

Kandasammy, Lloyd F. "A Brief History of Floods in Guyana." *Stabroek News* (Guyana), February 6, 2006. http://www.landofsixpeoples.com/news601/ns6021650.htm.

Kates, R. W., C. E. Colten, S. Laska, and S. P. Leatherman. "Reconstruction of New Orleans after Hurricane Katrina: A Research Perspective." *Proceedings of the National Academy of Sciences* 103, no. 40 (2006): 14653–60.

Kelman, Ilan, Justyna Orlowska, Himani Upadhyay, Robert Stojanov, Christian Webersik, Andrea C. Simonelli, David Procházka, and Daniel Němec. "Does Climate Change Influence People's Migration Decisions in Maldives?" *Climatic Change* 153, no. 1–2 (2019): 285–99.

Kench, Paul, Scott Nichol, and Roger McLean. "Comment on 'New Perspectives for the Future of the Maldives' by Mörner, N.A., et al." *Global and Planetary Change* 47 (2005): 67–69.

Kench, P. S., S. G. Smithers, R. F. McLean, and S. L. Nichol. "Holocene Reef Growth in the Maldives: Evidence of a Mid-Holocene Sea-Level Highstand in the Central Indian Ocean." *Geology* 37, no. 5 (2009): 455–58.

Keohane, Robert O., and Joseph S. Nye. "Globalization: What's New? What's Not? (and So What?)." *Foreign Policy* 118 (2000): 104–19.

Kimura, Shuhei. "When a Seawall Is Visible: Infrastructure and Obstruction in Post-Tsunami Reconstruction in Japan." *Science as Culture* 25, no. 1 (2016): 23–43.

Kincaid, Jamaica. *A Small Place.* New York: Farrar, Straus and Giroux, 1988.

Klein, Naomi. *The Battle for Paradise: Puerto Rico Takes on Disaster Capitalists.* Chicago: Haymarket, 2018.

——. *The Shock Doctrine: The Rise of Disaster Capitalism.* New York: Picador, 2007.

——. *This Changes Everything: Capitalism vs. the Climate.* New York: Simon & Schuster, 2015.

Kline, Benjamin. *First along the River: A Brief History of the US Environmental Movement.* New York: Arcada, 2000.

Koslov, Liz. "The Case for Retreat." *Public Culture* 28, no. 2 (2016): 359–87.

Kothari, Uma. "Authority and Expertise: The Professionalisation of International Development and the Ordering of Dissent." *Antipode* 37, no. 3 (2005): 425–46.

——. "Political Discourses of Climate Change and Migration: Resettlement Policies in the Maldives." *Geographical Journal* 180, no. 2 (2014): 130–40.

Kothari, Uma, and Alex Arnall. "Contestation over an Island Imaginary Landscape: The Management and Maintenance of Touristic Nature." *Environment and Planning A: Economy and Space* 49, no. 5 (2017): 980–98.

——. "Shifting Sands: The Rhythms and Temporalities of Island Sandscapes." *Geoforum* 108 (2020): 305–14.

Kothari, Uma, and Rorden Wilkinson. "Colonial Imaginaries and Postcolonial Transformations: Exiles, Bases, Beaches." *Third World Quarterly* 31, no. 8 (2010): 1395–412.

Kraus, Nicholas C., ed. *History and Heritage of Coastal Engineering: A Collection of Papers on the History of Coastal Engineering in Countries Hosting the International Coastal Engineering Conference 1950–1996.* New York: American Society of Civil Engineers, 1996.

Kril, Ray, and Rupert Roopnaraine, dir. *The Seawall, Tales of the Guyana Coast.* Guyana, 1992.

Larkin, Brian. *Signal and Noise: Media, Infrastructure, and Urban Culture in Nigeria.* Durham, NC: Duke University Press, 2008.

Le Guin, Ursula K. *The Dispossessed.* New York: Harper & Row, 1974.

Leitner, Helga, Eric Sheppard, Sophie Webber, and Emma Colven. "Globalizing Urban Resilience." *Urban Geography* 39, no. 8 (2018): 1276–84.

Levenstein, Steve. "Plush Tetrapods Turn Your Couch into a Coastline," http://inventorspot.com/articles/plush_tetrapods_turn_your_couch_coastline_24658. Accessed April 8, 2023.

Lewis, Arthur. *Labour in the West Indies: The Birth of a Workers' Movement.* London: Fabian Society, 1939.

Lewis, Simon L., and Mark A. Maslin. "Defining the Anthropocene." *Nature* 519, no. 7542 (2015): 171–80.

Lipset, Seymour Martin. *Political Man: The Social Bases of Politics.* New York: Doubleday, 1960.

MacKinnon, Danny, and Kate Driscoll Derickson. "From Resilience to Resourcefulness: A Critique of Resilience Policy and Activism." *Progress in Human Geography* 37, no. 2 (2012): 253–70.

Macy, Joanna. *Pass It On: Five Stories That Can Change the World.* Berkeley, CA: Parallax Press, 2010.

Mahoney, Clarence. *People of the Maldive Islands.* Himayatnagar, Hyderabad: Orient Blackswan, 2013.

Malm, Andreas. "Sea Wall Politics: Uneven and Combined Protection of the Nile Delta Coastline in the Face of Sea Level Rise." *Critical Sociology* 39, no. 6 (2013): 803–32.

Mangal, Jan. "National Oil Company for Guyana Would Be a Disaster." *Stabroek News* (Guyana), July 2, 2018.

Marino, Elizabeth K. *Fierce Climate, Sacred Ground: An Ethnography of Climate Change in Shishmaref Alaska.* Fairbanks: University of Alaska Press, 2015.

Marris, Emma. *Rambunctious Garden: Saving Nature in a Post-Wild World.* New York: Bloomsbury, 2011.

Martin, Cedriann J. "Annette Arjoon-Martins: 'The Grassroots People Are the Real Heroes.'" *Caribbean Beat*, 2011. https://www.caribbean-beat.com/issue-110/-grassroots-people-are-real-heroes#axzz7Z8KE1zNU.

Martinez, Andrea. "Gender and Development: Issues and Struggles of Third World Women." In *Introduction to International Development: Approaches, Actors, and Issues*, edited by Paul A. Haslam, Jessica Schafer, and Pierre Beaudet, 82–102. Oxford: Oxford University Press, 2009.

Martins, Dave. "Vreed en Hoop Then and Now." *Stabroek News* (Guyana), August 23, 2010. www.stabroeknews.com/2010/08/23/features/vreed-en-hoop-then-and-now/.

Matyas, David, and Mark Pelling. "Positioning Resilience for 2015: The Role of Resistance, Incremental Adjustment and Transformation in Disaster Risk Management Policy." *Disasters* 39, no. s1 (2014): s1–s18.

Maza, Antonio J., Rodolfo Silva, and Carlos Sanchez. "History of Coastal Engineering in Mexico." In *History and Heritage of Coastal Engineering: A Collection of Papers on the History of Coastal Engineering in Countries Hosting the International Coastal Engineering Conference 1950–1996*, edited by Nicholas C. Kraus, 375–89. New York: American Society of Civil Engineers, 1996.

McAdam, Jane. "'Disappearing States,' Statelessness and the Boundaries of International Law." *UNSW Law Research Paper*, no. 2010–2 (2010). https://ssrn.com/abstract=1539766.

McMichael, Philip. *Development and Social Change: A Global Perspective.* Los Angeles: Sage, 2017.

Mead, Allison A., and Michael T. Lee. "Sediment Exchange between Amazon Mudbanks and Shore-Fringing Mangroves in British Guiana." *International Journal of Marine Geology, Geochemistry and Geophysics* 208 (2004): 169–90.

Meriläinen, Eija. "The Dual Discourse of Urban Resilience: Robust City and Self-Organised Neighbourhoods." *Disasters* 44, no. 1 (2020): 125–51.

Milligan, Brett, "Tetrapods, Entropy and Excess," *Free Association Design*, September 18, 2010, https://freeassociationdesign.wordpress.com/2010/09/18/tetrapods-entropy-and-excess/.

Moosa, Ahdha, Khoa Do, and Emil Jonescu. "Design Response to Rising Sea Levels in the Maldives: A Study into Aquatic Architecture." *Frontiers of Architectural Research* 9, no. 3 (2020): 623–40.

Morita, Atsuro. "Infrastructuring Amphibious Space: The Interplay of Aquatic and Terrestrial Infrastructures in the Chao Phraya Delta in Thailand." *Science as Culture* 25, no. 1 (2016): 117–40.

Mörner, Nils-Axel. "Claim that Sea Level is Rising is a Total Fraud." *Executive Intelligence Review* 34, no. 25 (2007).

———. "Why the Maldives Aren't Sinking." *Spectator*, December 2, 2009. www.spectator.co.uk/article/why-the-maldives-aren-t-sinking.

Mörner, Nils-Axel, Michael Tooley, and Göran Possnert. "New Perspectives for the Future of the Maldives." *Global and Planetary Change* 40, no. 1 (2004): 177–82.

Moser, Susanne, Sara Meerow, James Arnott, and Emily Jack-Scott. "The Turbulent World of Resilience: Interpretations and Themes for Transdisciplinary Dialogue." *Climatic Change* 153, no. 1–2 (2019): 21–40.

Moshavi, Sharon. "Tourists Flock to a Sinking Paradise." *Bloomberg Businessweek*, September 17, 1995. http://businessweek.com/stories/1995-09-17/tourists-flock-to-a-sinking-paradise-dot-dot-dot-intl-edition.

Mukerji, Chandra. "Stewardship Politics and the Control of Wild Weather: Levees, Seawalls, and State Building in 17th-Century France." *Social Studies of Science* 37 (2007): 127–33.

Naseem, Azra, and Mohamed Mushfique. "The Long Road from Islam to Islamism: A Short History." *Dhivehi Sitee*, May 30, 2014. http://www.dhivehisitee.com/religion/islamism-maldives/.

National Development Strategy Secretariat. "Water Management and Flood Control." Ministry of Finance, Guyana, 1996. http://guyana.org/NDS/chap40.htm#2contents_A.

Naylor, Alexander K. "Island Morphology, Reef Resources, and Development Paths in the Maldives." *Progress in Physical Geography: Earth and Environment* 39, no. 6 (2015): 728–49.

Niemeyer, Hanz D., Hartmut Eiben, and Hans Rohde. "History and Heritage of German Coastal Engineering." In *History and Heritage of Coastal Engineering: A Collection of Papers on the History of Coastal Engineering in Countries Hosting the International Coastal Engineering Conference 1950–1996*, edited by Nicholas C. Kraus, 169–213. New York: American Society of Civil Engineers, 1996.

Nunn, Patrick D. "Understanding and Adapting to Sea-Level Rise." In *Global Environmental Issues*, edited by Frances Harris, 87–104. Oxford: Wiley-Blackwell, 2012.

"Obituary. Edward Case, 1842–1899." *Minutes and Proceedings of the Institution of Civil Engineers* 139 (1900): 374–76.

O'Conner, James. "Is Sustainable Capitalism Possible?" In *Is Capitalism Sustainable*, edited by Martin O'Conner, 152–75. New York: Guilford Press, 1994.

Octavianti, Thanti, and Katrina Charles. "Disaster Capitalism? Examining the Politicisation of Land Subsidence Crisis in Pushing Jakarta's Seawall Megaproject." *Water Alternatives* 11, no. 2 (2018): 394–420.

Ogden, Laura A. *Swamplife: People, Gators, and Mangroves Entangled in the Everglades*. Minneapolis: University of Minnesota Press, 2011.

Ortner, Sherry B. "Is Female to Male as Nature Is to Culture?" In *Women, Culture, and Society*, edited by M. Z. Rosaldo and L. Lamphere, 68–87. Stanford, CA: Stanford University Press, 1974.

Palmer, Rendel. "History of Coastal Engineering in Great Britain." In *History and Heritage of Coastal Engineering: A Collection of Papers on the History of Coastal Engineering in Countries Hosting the International Coastal Engineering Conference 1950–1996*, edited by Nicholas C. Kraus, 214–74. New York: American Society of Civil Engineers, 1996.

Paprocki, Kasia. "All That Is Solid Melts into the Bay: Anticipatory Ruination and Climate Change Adaptation." *Antipode* 51, no. 1 (2019): 295–315.

———. "The Climate Change of Your Desires: Climate Migration and Imaginaries of Urban and Rural Climate Futures." *Environment and Planning D: Society and Space* 38, no. 2 (2019): 248–66.

"Past Evolution of Politics: Sea Defences." Government of Guyana, 1996. Accessed August 8, 2020. http://www.guyana.org/NDS/chap40.htm#2contents_A.

Pauly, Daniel. "Anecdotes and the Shifting Baseline Syndrome of Fisheries." *Trends in Ecology & Evolution* 10, no. 10 (1995): 430.

Pellow, David Naguib. *Resisting Global Toxics: Transnational Movements for Environmental Justice*. Cambridge, MA: MIT Press, 2007.

Pilkey, Orrin H., and Rob Young. *The Rising Sea*. Washington, DC: Island Press, 2009.

Pliny the Elder. "Countries That Have No Trees." Translated by John Bostock and H. T. Riley. In *Natural History*, 339–40. London: Taylor and Francis, 1855.

Pyrard, François. *The Voyage of François Pyrard of Laval to the East Indies, the Maldives, the Moluccas and Brazil.* Vol. 3. Edited by Albert Gray and H. C. P. Bell. Cambridge: Cambridge University Press, 2010.

Rabe, Stephan G. *U.S. Intervention in British Guiana: A Cold War Story.* Chapel Hill: University of North Carolina Press, 2005.

Ranganathan, Malini, and Eve Bratman. "From Urban Resilience to Abolitionist Climate Justice in Washington, DC." *Antipode* 53, no. 1 (2021): 115–37.

Rice, Jennifer L., Joshua Long, and Anthony Levenda. "Against Climate Apartheid: Confronting the Persistent Legacies of Expendability for Climate Justice." *Environment and Planning E: Nature and Space* 5, no. 2 (2022): 625–45.

Rist, Gilbert. *The History of Development: From Western Origins to Global Faith.* 2nd ed. London: Zed Books, 2002.

Rist, L., A. Felton, M. Nyström, M. Troell, R. A. Sponseller, J. Bengtsson, H. Österblom, et al. "Applying Resilience Thinking to Production Ecosystems." *Ecosphere* 5, no. 6 (2014). http://dx.doi.org/10.1890/ES13-00330.1.

Robinson, Cedric. *Black Marxism: The Making of the Black Radical Tradition.* Chapel Hill: University of North Carolina Press, 1983.

Rochette, Louis Stanislas D'Arcy De La. "Guyana Coast." London: William Faden, 1783.

Rodney, Walter. *Guyanese Sugar Plantations in the Late Nineteenth Century: A Contemporary Description from the Argosy.* Georgetown, Guyana: Release Publishers, 1979.

———. *A History of the Guyanese Working People, 1881–1905.* Baltimore: Johns Hopkins University Press, 1981.

———. "The Struggle Goes On: A Speech Made in September 1979." History Is a Weapon: Working People's Alliance, 1984. https://www.historyisaweapon.com/defcon1/rodnstrugoe.html.

Rodway, James, and James Graham Cruickshank. *The Story of Georgetown: Revised from a Series of Articles in the "Argosy" 1903.* Demerara, Guyana: Argosy, 1920.

Rogerson, Jayne M. "'Kicking Sand in the Face of Apartheid': Segregated Beaches in South Africa." *Bulletin of Geography* 35, no. 35 (2017): 93–110.

Romero-Frías, Xavier. *The Maldive Islanders: A Study of the Popular Culture of an Ancient Ocean Kingdom.* Barcelona: Nova Ethnographia Indica, 2003.

Rostow, Walt Whitman. *The Stages of Economic Growth: A Non-Communist Manifesto.* Cambridge: Cambridge University Press, 1960.

Rush, Elizabeth. *Rising: Dispatches from the New American Shore.* Minneapolis: Milkweed Editions, 2018.

Russo, Antonio Paolo. "The 'Vicious Circle' of Tourism Development in Heritage Cities." *Annals of Tourism Research* 29, no. 1 (2002): 165–82.

Sachs, Carolyn. *Women Working in the Environment.* Washington, DC: Taylor & Francis, 1997.

Sachs, Wolfgang, ed. *The Development Dictionary: A Guide to Knowledge as Power.* 2nd ed. New York: Zed Books, 2010.

Scott, James C. *Seeing Like a State: How Certain Schemes to Improve the Human Condition Have Failed.* New Haven, CT: Yale University Press, 1999.

"Sea Wall 'Saves Maldives Capital,'" *BBC News*, January 10, 2005. http://news.bbc.co.uk/2/hi/south_asia/4161491.stm.

Shareef, Ali. "A Tribute to the Late Ameed Didi." *Addu Diary* [Blog], 2010. http://addudiary.blogspot.com/2010/08/tribute-to-late-ameen-didi.html.

Shearer, Christine. *Kivalina: A Climate Change Story.* Chicago: Haymarket, 2011.

Sheller, Mimi. *Island Futures: Caribbean Survival in the Anthropocene.* Durham, NC: Duke University Press, 2020.

———. "The Islanding Effect: Post-Disaster Mobility Systems and Humanitarian Logistics in Haiti." *Cultural Geographies* 20, no. 2 (2013): 185–204.

———. "Theorising Mobility Justice." *Tempo Social* 30 (2018): 17–34.

Shenk, Jon, dir. *The Island President.* First Run Features, 2012.

Shiva, Vandana. "Ecological Balance in an Era of Globalization." In *The Globalization Reader*, edited by Frank J. Lechner and John Boli, 566–74. London: Routledge, 2000.

———. *Staying Alive: Women, Ecology and Development.* London: Zed Books, 1989.

Shuto, Nobuo, and Koji Fujima. "A Short History of Tsunami Research in Japan." *Proceedings of the Japan Academy, Series B Physical and Biological Sciences* 85, no. 8 (2009): 267–75.

Siders, A. R. "Social Justice Implications of US Managed Retreat Buyout Programs." *Climatic Change* 152, no. 2 (2019): 239–57.

Sims, Benjamin. "Things Fall Apart: Disaster, Infrastructure, and Risk." *Social Studies of Science* 37, no. 1 (2007): 93–95.

Smith, Neil. *Uneven Development: Nature, Capital, and the Production of Space.* 3rd ed. Athens: University of Georgia Press, 2008.

So, Alvin Y. *Social Change and Development: Modernization, Dependency, and World-Systems Theories.* Newbury Park, CA: Sage, 1990.

Sogge, David. "Multilateral Actors in Development." In *Introduction to International Development: Approaches, Actors, and Issues*, edited by Alexander Paul Haslam, Jessica Schafer, and Pierre Beaudet, 169–90. Oxford: Oxford University Press, 2009.

Sørensen, Torben, Jorgen Fredsøe, and Per Roed Jakobsen. "History of Coastal Engineering in Denmark." In *History and Heritage of Coastal Engineering: A Collection of Papers on the History of Coastal Engineering in Countries Hosting the International Coastal Engineering Conference 1950–1996*, edited by Nicholas C. Kraus, 103–41. New York: American Society of Civil Engineers, 1996.

Sovacool, Benjamin K. "Expert Views of Climate Change Adaptation in the Maldives." *Climatic Change* 114, no. 2 (2012): 295–300.

———. "Hard and Soft Paths for Climate Change Adaptation." *Climate Policy* 11, no. 4 (2011): 1177–83.

———. "Perceptions of Climate Change Risks and Resilient Island Planning in the Maldives." *Mitigation and Adaptation Strategies for Global Change* 17, no. 7 (2012): 731–52.

Star, Susan Leigh. "The Ethnography of Infrastructure." *American Behavioral Scientist* 43, no. 3 (1999): 377–91.

Star, Susan Leigh, and James R. Griesemer. "Institutional Ecology, 'Translations' and Boundary Objects: Amateurs and Professionals in Berkeley's Museum of Vertebrate Zoology, 1907–39." *Social Studies of Science* 19, no. 3 (1989): 387–420.

Steinberg, Philip E., Elizabeth Nyman, and Mauro J. Caraccioli. "Atlas Swam: Freedom, Capital, and Floating Sovereignties in the Seasteading Vision." *Antipode* 44, no. 4 (2012): 1532–50.

Stoler, Ann Laura, ed. *Imperial Debris: On Ruins and Ruination*. Durham, NC: Duke University Press, 2013.

Sundberg, Adam. "Molluscan Explosion: The Dutch Shipworm Epidemic of the 1730s." *Arcadia: Explorations in Environmental History*, no. 14 (2015). https://doi.org/10.5282/rcc/7307.

"The Suspension of the British Guiana Constitution—1953 (Declassified British Documents)." Updated August 2004, accessed August 8, 2020. http://www.guyana.org/govt/declassified_british_documents_1953.html.

Sze, Julie. *Fantasy Islands: Chinese Dreams and Ecological Fears in an Age of Climate Crisis*. Berkeley: University of California Press, 2015.

Tsing, Anna Lowenhaupt. "The Buck, the Bull, and the Dream of the Stag: Some Unexpected Weeds of the Anthropocene." *Suomen Antropologi: Journal of the Finnish Anthropological Society* 42, no. 1 (2017): 3–21.

———. *Friction: An Ethnography of Global Connection*. Princeton, NJ: Princeton University Press, 2004.

UN Human Rights Council. *Report of the Special Rapporteur on the Human Rights of Internally Displaced Persons, Addendum: Mission to Maldives*. January 30, 2012. https://www.refworld.org/docid/4f3935d32.html.

US Department of Arts and Culture. "Creative Placemaking, Placekeeping, and Cultural Strategies to Resist Displacement." USDAC Citizen Artist Salon, March 8, 2016. https://actionnetwork.org/forms/watch-the-creative-placekeeping-citizen-artist-salon.

US Department of State. "Progress Report to the 303 on Guyana," by Richard Lehman. Washington, DC, June 17, 1969. https://static.history.state.gov/frus/frus1969-76ve10/pdf/d366.pdf.

Van Veen, Johan. *Dredge Drain Reclaim: The Art of a Nation*. 4th ed. Dordrecht: Springer, 1955.

Vaughn, Sarah E. "Disappearing Mangroves: The Epistemic Politics of Climate Adaptation in Guyana." *Cultural Anthropology* 32, no. 2 (2017): 242–68.

Wave of Change—Maldives. March 2005. Video, 24:58. https://www.youtube.com/watch?v=GYT-XSnhw_g.

Webber, Sophie. "Performative Vulnerability: Climate Change Adaptation Policies and Financing in Kiribati." *Environment and Planning A: Economy and Space* 45, no. 11 (2013): 2717–33.

Weichselgartner, Juergen, and Ilan Kelman. "Geographies of Resilience: Challenges and Opportunities of a Descriptive Concept." *Progress in Human Geography* 39, no. 3 (2015): 249–67.

Weisman, Alan. *The World without Us*. New York: St. Martin's, 2007.

Wesselink, Anna J., Wiebe E. Bijker, Huib J. de Vriend, and Maarten S. Krol. "Dutch Dealings with the Delta." *Nature and Culture* 2, no. 2 (2007): 188–209.

Widener, Patricia. "Coastal People Dispute Offshore Oil Exploration: Toward a Study of Embedded Seascapes, Submersible Knowledge, Sacrifice, and Marine Justice." *Environmental Sociology* 4, no. 4 (2018): 405–18.

Williams, Michael. *The Draining of the Somerset Levels*. Cambridge: Cambridge University Press, 1970.

Wilshusen, Peter R. "Exploring the Political Contours of Conservation." In *Contested Nature: Promoting International Biodiversity with Social Justice in the Twenty-First Century*, edited by Steven R. Brechin, Peter R. Wilshusen, Crystal L. Fortwangler, and Patrick C. West, 41–58. New York: State University of New York Press, 2003.

Winner, Langdon. "Do Artifacts Have Politics?" *Daedalus* 109, no. 1 (1980): 121–36.

Wodehouse, Philip Edmond. "Correspondence Respecting the Sea Wall at Demerara." In *Accounts and Papers of the House of Commons 1854–55*, 33–34. Great Britain: House of Commons.

World Bank. "Appraisal of Second Sea Defense Project (Georgetown Urban Protective Works) Guyana." Washington, DC, 1971. http://documents1.worldbank.org/curated/en/574601468250853198/text/multi-page.txt.

——. "Appraisal Report: Sea Defense Project Guyana." International Bank for Reconstruction and Development, Washington, DC, 1968. http://documents1.worldbank.org/curated/en/291571468275676329/text/multiopage.txt.

World Commission on Environment and Development (WCED). *Our Common Future*. Oxford: Oxford University Press, 1987.

Zavestoski, Stephen. "The Struggle for Justice in Bhopal: A New/Old Breed of Transnational Social Movement." *Global Social Policy* 9, no. 3 (2009): 383–407.

Index

Founded in 1893,

UNIVERSITY OF CALIFORNIA PRESS
publishes bold, progressive books and journals on topics in the arts, humanities, social sciences, and natural sciences—with a focus on social justice issues—that inspire thought and action among readers worldwide.

The UC PRESS FOUNDATION
raises funds to uphold the press's vital role as an independent, nonprofit publisher, and receives philanthropic support from a wide range of individuals and institutions—and from committed readers like you. To learn more, visit ucpress.edu/supportus.